Materials for Energy Conversion and Storage

Development of new energy-related materials is essential in addressing future energy demands. *Materials for Energy Conversion and Storage* focuses on the materials science related to energy conversion and energy storage technologies. It covers the principles of prospective energy technologies and their relationship to the performance of energy devices.

- Covers fundamental principles of energy conversion and storage
- Discusses materials selection, design, and performance tradeoffs
- Details electrochemical cell construction and testing methodologies
- Explores sustainable development of energy devices
- Features case studies

Aimed at readers in materials, electrical, and energy engineering, this book provides readers with a deep understanding of the role of materials in developing sustainable energy devices.

Hieng Kiat Jun obtained his MEngSci and PhD from the University of Malaya, Malaysia. He is an Associate Professor in the Department of Mechanical and Materials Engineering, Universiti Tunku Abdul Rahman, Malaysia. He conducts experimental research on nanosized energy devices in solar cells and electrochemical energy storage systems. His main topics of interest include dye- and quantum dot-sensitized solar cells, organic solar cells, perovskite solar cells, electrolyte characterization, radiative cooling, and thermoelectric devices.

Foo Wah Low is an Assistant Professor in the Department of Electrical and Electronics Engineering at Universiti Tunku Abdul Rahman, Malaysia. He received his PhD in nanomaterials and nanoscience from Universiti Malaya. His main research interests are in the areas of graphene nanomaterials incorporated with metal oxide for photovoltaic cell application, especially in dye-sensitized solar cells and perovskites for photovoltaic cells.

Emerging Materials and Technologies
Series Editor: Boris I. Kharissov

The *Emerging Materials and Technologies* series is devoted to highlighting publications centered on emerging advanced materials and novel technologies. Attention is paid to those newly discovered or applied materials with potential to solve pressing societal problems and improve quality of life, corresponding to environmental protection, medicine, communications, energy, transportation, advanced manufacturing, and related areas.

The series takes into account that, under present strong demands for energy, material, and cost savings, as well as heavy contamination problems and worldwide pandemic conditions, the area of emerging materials and related scalable technologies is a highly interdisciplinary field, with the need for researchers, professionals, and academics across the spectrum of engineering and technological disciplines. The main objective of this book series is to attract more attention to these materials and technologies and invite conversation among the international R&D community.

Smart Micro- and Nanomaterials for Pharmaceutical Applications
Edited by Ajit Behera, Arpan Kumar Nayak, Ranjan K. Mohapatra, and Ali Ahmed Rabaan

Friction Stir-Spot Welding
Metallurgical, Mechanical and Tribological Properties
Edited by Jeyaprakash Natarajan and K. Anton Savio Lewise

Phase Change Materials for Thermal Energy Management and Storage
Fundamentals and Applications
Edited by Hafiz Muhammad Ali

Nanofluids
Fundamentals, Applications, and Challenges
Shriram S. Sonawane and Parag P. Thakur

MXenes
From Research to Emerging Applications
Edited by Subhendu Chakroborty

Biodegradable Polymers, Blends and Biocomposites
Trends and Applications
Edited by A. Arun, Kunyu Zhang, Sudhakar Muniyasamy and Rathinam Raja

For more information about this series, please visit: www.routledge.com/
Emerging-Materials-and-Technologies/book-series/CRCEMT

Materials for Energy Conversion and Storage

Edited by
Hieng Kiat Jun and Foo Wah Low

CRC Press
Taylor & Francis Group
Boca Raton London New York

CRC Press is an imprint of the
Taylor & Francis Group, an **Informa** business

Designed cover image: © 2025 Hieng Kiat Jun

First edition published 2025
by CRC Press
2385 NW Executive Center Drive, Suite 320, Boca Raton FL 33431

and by CRC Press
4 Park Square, Milton Park, Abingdon, Oxon, OX14 4RN

CRC Press is an imprint of Taylor & Francis Group, LLC

© 2025 selection and editorial matter, Hieng Kiat Jun and Foo Wah Low; individual chapters, the contributors

ISBN: 978-1-032-32311-4 (hbk)
ISBN: 978-1-032-32312-1 (pbk)
ISBN: 978-1-003-31442-4 (ebk)

DOI: 10.1201/9781003314424

Typeset in Times
by codeMantra

Contents

Preface...ix
Contributors xi

Chapter 1   Introduction: Energy and Materials 1

Hieng Kiat Jun

1.1   Energy Issue 1
1.2   Energy Transition 2
1.3   Energy Conversion 4
1.4   Materials and Energy 6
1.5   Trends in Materials Research 8
1.6   Conclusion 10
References 10

Chapter 2   Photovoltaic: Fundamental of Energy Conversion 12

Hieng Kiat Jun and Foo Wah Low

2.1   Introduction and Brief History 12
2.2   Basic Characteristics and Characterization of
        Photovoltaics... 13
2.3   Photocurrent Generation and Photovoltage 14
2.4   Charge Separation, Recombination, and Efficiency.... 16
2.5   Conclusion 18
Acknowledgment... 18
References 19

Chapter 3   Photovoltaic: Graphene Materials and Their Effects in
                 Perovskite Solar Cells.. 20

*Foo Wah Low, Muhammad Kashif, Mohammad Aminul Islam,
Hieng Kiat Jun, Cheen Sean Oon, and Chun Hong Voon*

3.1   Introduction 20
3.2   Crystalline Si-Based Photovoltaic 21
3.3   III–V Semiconductors Multijunction-Based Photovoltaic....... 22
3.4   Thin-Film Photovoltaic 23
3.5   Nanocomposite Solar Cells 24
        3.5.1   Perovskites Solar Cells (PSCs)................... 24
        3.5.2   Structure of PSCs 26
        3.5.3   Working Principle of PSCs 26
        3.5.4   Electron Transport Layer (ETL)................. 27
3.6   Graphene.. 28

3.6.1   Physical/Mechanical Properties of Graphene 29
3.6.2   Electronic Properties of Graphene 29
3.6.3   Graphene-Based Electron Transport Layer 30
3.6.4   Defects of Graphene 31
3.6.5   Type of Defects in Graphene 32
3.6.6   Solution for Graphene Defects 34
3.7   Conclusion 37
Acknowledgment 37
References 37

Chapter 4   Fuel Cells: Fundamental and Applications 39

Chiam-Wen Liew, Son Qian Liew, and Hieng Kiat Jun

4.1   Introduction to Fuel Cell 39
4.2   Principles of Fuel Cell Operation 39
       4.2.1   Mechanism of PEMFCs 39
       4.2.2   Thermodynamics Principles of PEMFCs 40
4.3   Category of Fuel Cells 43
       4.3.1   Alkaline Fuel Cells (AFCs) 43
       4.3.2   Solid Oxide Fuel Cells (SOFCs) 45
       4.3.3   Phosphoric Acid Fuel Cells (PAFCs) 47
       4.3.4   Molten Carbonate Fuel Cells (MCFCs) 49
       4.3.5   Direct Methanol Fuel Cells (DMFCs) 49
       4.3.6   Polymer Electrolyte Membrane Fuel Cells
                (PEMFCs) 51
       4.3.7   Biofuel Cells 52
4.4   Advantages and Disadvantages of PEMFCs 52
4.5   Applications of Fuel Cells 53
       4.5.1   Stationary Power Generation Plants 53
       4.5.2   Automotive Industry 55
4.6   Conclusions 57
Acknowledgment 58
References 58

Chapter 5   Solid-State Polymer Electrolytes for Fuel Cells Application 61

Chiam-Wen Liew, Son Qian Liew, and Hieng Kiat Jun

5.1   Introduction 61
5.2   Development of Electrolytes 62
       5.2.1   Liquid Electrolytes 62
       5.2.2   Solid Electrolytes 62
       5.2.3   Solid-State Polymer Electrolytes 63
       5.2.4   General Requirements of Solid-State
                Polymer Electrolytes for Electrochemical
                Devices Applications 67

5.3   Description of Ionic Conduction 68
        5.3.1   Mechanism of Ionic Conduction 68
        5.3.2   Requirements of Ionic Conductivity 70
        5.3.3   Governing Parameters of Ionic Conduction 70
        5.3.4   Methods of Enhancing Ionic Conduction
                Mechanism 71
5.4   Future Aspects 85
        5.4.1   Liquid Crystal Polymer Electrolytes (LCPEs) 85
        5.4.2   Liquid Crystals-Embedded Polymer Electrolytes 86
5.5   Conclusions 87
Acknowledgment 87
References 87

Chapter 6   Next-Generation Supercapacitors with Sustainably Processed
                 Carbon Quantum Dots 97

Grishika Arora, Tejas Sharma, Foo Wah Low, Chai Yan Ng,
Pramod K. Singh, Chiam-Wen Liew, and Hieng Kiat Jun

6.1   Introduction 97
6.2   General Characteristics of Supercapacitors and EDLCs 98
6.3   Working Principle and Classification of Supercapacitors 99
        6.3.1   Electric Double-Layer Capacitors (EDLCs) 103
        6.3.2   Pseudocapacitors 104
        6.3.3   Hybrid Capacitors 105
6.4   Carbon Quantum Dots (CQDs) 106
        6.4.1   Properties of CQDs 107
        6.4.2   Synthesis of CQDs 109
        6.4.3   Synthesis of CQDs from Biomass Wastes 114
6.5   Application of "Green" CQDs in Supercapacitors 116
6.6   Conclusions and Perspectives 120
Acknowledgment 122
References 122

Chapter 7   Tin-Based Anodes for Next-Generation Lithium-Ion Batteries 127

L. P. Teo, M. H. Buraidah, and A. K. Arof

7.1   Introduction 127
7.2   Tin-Based Alloys 129
        7.2.1   Tin-Based Oxides 130
        7.2.2   Tin (II) Oxide (SnO) 130
        7.2.3   Tin (IV) Oxide (SnO$_2$) 132
        7.2.4   SnO$_2$ Nanowires 134
        7.2.5   SnO$_2$ Nanorods 135
        7.2.6   SnO$_2$ Nanotubes 136
        7.2.7   SnO$_2$ Nanosheets 136
        7.2.8   SnO$_2$/Carbon Composites 137

7.3   Ternary Tin Oxides 138
       7.3.1   Lithium Tin Oxide or Lithium
               Stannate (Li$_2$SnO$_3$) 139
       7.3.2   Calcium Metastannate (CaSnO$_3$)........................... 142
       7.3.3   Strontium Metastannate (SrSnO$_3$) 142
       7.3.4   Barium Metastannate (BaSnO$_3$)............................. 145
       7.3.5   Cobalt Metastannate (CoSnO$_3$) 145
       7.3.6   Cadmium Metastannate (CdSnO$_3$)........................ 147
       7.3.7   Zinc Metastannate (ZnSnO$_3$)................................. 148
       7.3.8   Zinc Stannate (Zn$_2$SnO$_4$).................................. 148
7.4   Other Stannates 149
7.5   Summary 151
Acknowledgments 151
References 151

Chapter 8   Sustainable Materials for Energy Conversion System 159

            Grishika Arora and Hieng Kiat Jun

8.1   Introduction 159
8.2   Energy Conversion and Storage Systems 163
       8.2.1   Solar Energy.. 163
       8.2.2   Bioenergy Technologies 164
       8.2.3   Ocean Energy Technology 165
8.3   Life Cycle Analysis 166
       8.3.1   Life Cycle Analysis of Batteries............................. 167
       8.3.2   Life Cycle Analysis of Solar Cells 171
8.4   Recycling as Sustainable Usage of Materials...................... 177
       8.4.1   Acquisition, Preliminary Processing, and
               Finishing Off 178
       8.4.2   Leveraging the Recycling Process 178
       8.4.3   Future Recycling Concerns to Be Addressed.......... 179
8.5   New Options of Sustainable Usage in
       Energy Conversion 179
       8.5.1   Fuel Cells: The Conversion of Hydrogen 179
       8.5.2   Battery Rechargeables.. 180
8.6   Conclusion 181
Acknowledgement 181
References 182

Index... 187

Preface

Embarking on a profound exploration of *Materials for Energy Conversion and Storage*, this volume delves into various chapters, each dedicated to unraveling the complexities and advancements within specific domains of energy conversion technologies. In this 21st century, where rapid technological developments and the demands of our digitalized lifestyles rely heavily on electricity, the need for sustainable energy sources becomes increasingly imperative. Fossil fuels, our traditional mainstay, are finite and environmentally impactful, prompting a shift toward alternative, clean energy solutions.

This book's introductory chapter sets the stage by providing fundamental insights into the principles of energy conversion, in general and in photovoltaics, in the following chapter. Readers are guided through the intricate processes involved in harnessing solar energy for electricity, gaining a foundational understanding of various photovoltaic technologies, including silicon-based and thin films. The journey continues with the third chapter, which zooms into the captivating realm of graphene materials and their profound effects on perovskite solar cells (PSCs). Exploring the synergistic relationship between graphene and PSCs, this chapter illuminates the potential for pushing the efficiency boundaries of solar energy conversion.

Shifting the focus to fuel cells, the fourth chapter unfolds the fundamental principles and diverse applications of this technology. From transportation to stationary power generation, readers are introduced to the versatility of fuel cells and their role in providing clean and efficient energy solutions. Building upon this foundation, the subsequent chapter narrows its focus to delve into the intricate world of solid polymer electrolytes and their application in fuel cells. By unraveling the complexities of these materials, readers gain insights into the advancements driving fuel cell technology toward enhanced efficiency, durability, and real-world applicability.

The exploration extends to supercapacitors in the sixth chapter, with a specific emphasis on sustainably processed carbon quantum dots. This section addresses the pivotal role of these innovative materials in developing supercapacitors with improved performance and environmental sustainability. Continuing the journey into energy storage, the seventh chapter dives into the world of tin-based anodes, a crucial component in the development of next-generation lithium-ion batteries. Readers gain valuable insights into the potential of these anodes to enhance the energy density, lifespan, and safety of lithium-ion batteries.

The final chapter casts a wide net, encompassing sustainable materials for energy conversion systems. Synthesizing key learnings from preceding chapters, this section expands the discussion to address the broader landscape of materials contributing to sustainable energy conversion.

In essence, *Materials for Energy Conversion and Storage* presents a collective exploration of cutting-edge materials shaping the future of energy conversion. Each chapter not only delves into the technical intricacies of its respective fields but also emphasizes the broader goal of achieving efficiency, durability, and environmental sustainability in the quest for innovative energy solutions. As we navigate through these chapters, a holistic understanding of the intricate interplay between materials and energy conversion emerges, paving the way for a greener and more sustainable energy future.

Contributors

A. K. Arof
Universiti Malaya
Kuala Lumpur, Malaysia

Grishika Arora
Universiti Tunku Abdul Rahman
Kajang, Malaysia

M. H. Buraidah
Universiti Malaya
Kuala Lumpur, Malaysia

Mohammad Aminul Islam
Universiti Malaya
Kuala Lumpur, Malaysia

Hieng Kiat Jun
Universiti Tunku Abdul Rahman
Kajang, Malaysia

Muhammad Kashif
Tianjin University
Tianjin, China

Chiam-Wen Liew
Tunku Abdul Rahman University
    of Management and Technology
    (TARUMT)
Kuala Lumpur, Malaysia

Son Qian Liew
Universiti Tunku Abdul Rahman
Kajang, Malaysia

Foo Wah Low
Universiti Tunku Abdul Rahman
Kajang, Malaysia

Chai Yan Ng
Universiti Tunku Abdul Rahman
Kajang, Malaysia

Cheen Sean Oon
Monash University Malaysia
Bandar Sunway, Malaysia

Tejas Sharma
Universiti Tunku Abdul Rahman
Kajang, Malaysia

Pramod K. Singh
Sharda University
Greater Noida, India

L. P. Teo
Tunku Abdul Rahman University
    of Management and Technology
    (TARUMT)
Kuala Lumpur, Malaysia

Chun Hong Voon
University Malaysia Perlis
Perlis, Malaysia

1 Introduction
Energy and Materials

Hieng Kiat Jun

1.1 ENERGY ISSUE

Energy is the backbone of modern societies, powering everything from transportation and communication to agriculture and manufacturing. However, the world is facing an energy crisis due to various factors such as rising global population, urbanization, and the demand for better living standards. The issue of energy is multifaceted, and it encompasses various aspects of human activity, including environmental sustainability, economic development, and social justice (Butler et al., 2018). In this chapter, the energy issue and its impact on the world are briefly discussed.

One of the main challenges posed by the energy issue is the finite nature of fossil fuels, which currently supply the majority of the world's energy needs. Fossil fuels such as coal, oil, and gas are finite resources that will eventually run out, and their extraction and consumption contribute significantly to greenhouse gas emissions that cause climate change (Scovronick et al., 2019). This means that there is a pressing need to find alternative sources of energy that are sustainable and do not contribute to global warming (Figure 1.1).

Renewable energy sources such as solar, wind, hydropower, and geothermal are increasingly being viewed as viable alternatives to fossil fuels. These sources of energy are abundant, clean, and renewable. They have the potential to reduce greenhouse gas emissions while meeting the world's energy needs. However, the adoption of renewable energy sources is often hampered by the high cost of technology and the lack of infrastructure to support their use. Fortunately, the cost of renewable energy has been reducing from year to year due to policy and reduced manufacturing costs (IRENA, 2023).

Another aspect of the energy issue is the need for energy security. Many countries rely heavily on imported fossil fuels to meet their energy needs, which can lead to vulnerability and instability in the event of supply disruptions or price shocks. This has led to the exploration of domestic sources of energy such as shale gas, tar sands, and offshore oil and gas reserves. However, the extraction of these resources can be environmentally damaging, and there are concerns about the long-term sustainability of these sources of energy (Stevens, 2012).

The energy issue is also closely tied to economic development. The availability of affordable and reliable energy is essential for economic growth, but energy prices can also have a significant impact on economic stability. High energy prices can lead to inflation and slow economic growth, while low energy prices can discourage investment in renewable energy and reduce incentives for energy efficiency. This means

DOI: 10.1201/9781003314424-1

FIGURE 1.1 Reduction of greenhouse gas emission is paramount in reversing climate change.

that there is a delicate balance between energy prices and economic growth, and policymakers must carefully consider the impact of energy policies on the economy.

Besides, the energy issue also has social justice implications. Access to energy is a basic human right, and yet over one billion people worldwide lack access to electricity (Jenkins, 2018). This lack of access to energy can have a profound impact on health, education, and economic opportunities. Addressing energy poverty is therefore an important aspect of the energy issue, and it requires a concerted effort from governments, international organizations, and the private sector.

The energy issue is a complex and multifaceted problem that requires a comprehensive approach. The world needs to transition to a more sustainable energy system that relies on renewable energy sources and reduces greenhouse gas emissions. This transition will require significant investments in technology, infrastructure, and policy, as well as a shift in societal attitudes toward energy consumption and conservation. Addressing the energy issue will also require a commitment to social justice, ensuring that access to energy is available to all. One of the policy approaches is the energy transition policy (Figure 1.2).

1.2 ENERGY TRANSITION

Energy transition is the process of moving from traditional, non-renewable energy sources, such as fossil fuels, to cleaner and more sustainable sources of energy. The idea of energy transition is based on the need to reduce greenhouse gas emissions and mitigate the effects of climate change. What is more important is the impact of energy

FIGURE 1.2 Renewable energy like wind energy can be one of the solutions for energy issue.

transition that affects various aspects of society, including the economy, environment, and public health (Kabeyi and Olanrewaju, 2022). Meeting the needs of energy transition requires a concerted effort from governments, businesses, and individuals.

The main principle of energy transition is to shift from high-carbon energy sources to low-carbon or carbon-free energy sources (Khan et al., 2022). The primary goal is to reduce greenhouse gas emissions and mitigate the effects of climate change. This requires the use of renewable energy sources such as wind, solar, and hydroelectric power, as well as energy efficiency measures such as insulation and efficient appliances. Energy transition also involves the electrification of transportation, such as the use of electric vehicles, to reduce reliance on fossil fuels.

Energy transition offers numerous benefits. The most anticipated benefits are the reduction of greenhouse gas emissions and the improvement of air quality. These can have positive effects on public health. It can also stimulate economic growth and create new job opportunities in the renewable energy sector. However, energy transition can also induce negative impacts, such as job losses in the fossil fuel industry and higher costs of energy for consumers.

One of the main requirements for energy transition is the development and implementation of policies and regulations that support the transition to renewable energy sources (Steinbacher and Röhrkasten, 2019; Wang et al., 2021). Governments must provide incentives for the development and use of renewable energy, such as subsidies and tax credits. They can also impose regulations on greenhouse gas emissions and energy efficiency standards. Businesses also have a role to play, by investing in renewable energy technologies and adopting energy-efficient practices. Finally, individuals can contribute by making conscious choices in their energy consumption, such as using public transportation and reducing energy consumption at home.

FIGURE 1.3   Transitioning from fossil fuel–based energy to renewable energy is crucial for addressing climate change and pollution issues.

Another criterion for energy transition is the development of new and improved technologies for renewable energy production and energy storage. Renewable energy technologies, such as solar and wind power, have made significant progress in recent years but still face challenges such as intermittency and storage. Developing new technologies for energy storage, such as batteries and hydrogen fuel cells, can help overcome these challenges and increase the reliability and effectiveness of renewable energy sources. Furthermore, education and awareness are critical for achieving a successful energy transition. Educating the public about the benefits and challenges of renewable energy sources can help build support for energy transition policies and encourage individuals to make conscious choices in their energy consumption. Awareness campaigns can also help debunk myths and misconceptions about renewable energy, such as the belief that it is too expensive or unreliable. Understanding the principles of energy conversion is the beginning of the innovation of energy devices (Figure 1.3).

1.3   ENERGY CONVERSION

Energy conversion is the process of transforming one form of energy into another, such as converting solar energy into electricity or converting heat into mechanical work. The principle of energy conversion is based on the laws of thermodynamics, which govern the behavior of energy in physical systems (Lostaglio et al., 2014). The theory of energy conversion involves the study of the mechanisms and processes involved in energy transformation.

The First Law of thermodynamics states that energy cannot be created or destroyed but can only be transformed from one form to another. This law is also known as the law of conservation of energy. It signifies that the total amount of energy in a closed system remains constant and that energy cannot be created or destroyed within that system. Energy can only be transferred from one object or system to another or converted from one form to another.

The Second Law of thermodynamics states that the total entropy of a closed system always increases over time. Entropy is a measure of the disorder or randomness of a system. The Second Law implies that in any energy conversion process, some energy will inevitably be lost as unusable heat energy, and the overall efficiency of the process will be less than 100%. Therefore, energy conversion processes are always associated with some degree of inefficiency.

Energy conversion is an essential concept in physics and engineering and has numerous practical applications in various fields, including power generation, transportation, and electronics. There are various types of energy conversion processes, such as mechanical-to-electrical energy conversion, and photovoltaic energy conversion (Bejan, 2016).

In a mechanical-to-electrical energy conversion system, the process involves converting mechanical energy, such as the rotation of a wind turbine or the movement of a hydroelectric dam turbine, into electrical energy. This process involves the use of generators, which convert mechanical energy into electrical energy using the principle of electromagnetic induction. On the other hand, thermal-to-electrical energy conversion is a process of converting heat energy into electrical energy. An example of this system is a thermoelectric generator. This process involves the use of a thermocouple, which generates an electrical voltage when there is a temperature difference between two junctions.

For the conversion of chemical energy to electrical energy, the chemical energy stored in the system, such as batteries, is converted into electrical energy. This process involves a chemical reaction that generates an electrical potential difference between two electrodes. Electricity can also be generated from solar energy which is known as photovoltaic energy conversion. This process involves the conversion of solar energy into electrical energy, as found in a solar cell. Solar cell devices are usually made up of semiconductor materials, which absorb photons from sunlight, generating an electrical voltage in return.

The produced electricity is usually used to produce useful work or applications. In this case, electrical energy is converted to mechanical energy like in an electric motor. This process utilizes electromagnets to generate a magnetic field, inducing rotation in a conductor carrying an electric current through their interaction. In all the energy conversion systems, the amount of energy output produced determines the efficiency of the systems.

The efficiency of energy conversion processes is a crucial factor in determining their practical usefulness. It is defined as the ratio of the useful energy output to the energy input. For example, the efficiency of a wind turbine is defined as the ratio of the electrical power output to the wind power input. The efficiency of energy conversion processes is limited by various factors, including the laws of thermodynamics, the properties of the materials used in the conversion process, and the design of the conversion system (Nevskaya et al., 2023).

In recent years, there has been increasing interest in developing new and more efficient energy conversion technologies to address the growing demand for sustainable

FIGURE 1.4   Solar panel involves in converting solar energy to electric energy.

energy sources. One area of research is the development of materials with improved properties for energy conversion processes. For example, the development of new semiconductor materials with high efficiency and low cost is crucial for improving the efficiency of photovoltaic cells. Researchers are also exploring the use of nanomaterials and novel structures for improving the efficiency of energy conversion processes (Figure 1.4).

1.4   MATERIALS AND ENERGY

Materials science and energy are intricately linked, with materials playing a crucial role in the development of sustainable energy technologies. Materials science is a multidisciplinary field that encompasses the study of the structure, properties, and behavior of materials. It is a rapidly evolving field that has led to significant advancements in various areas like electronics, medicine, and energy.

We all acknowledge that energy is the driving force behind modern civilization. As the world continues to seek sustainable energy sources, the focus has shifted toward developing materials that can improve the efficiency, durability, and performance of energy systems. The use of suitable materials has enabled the development of new energy technologies in solar cells, wind turbines, and electric vehicles, to name a few. Indeed, materials science and engineering have played a vital role in the optimization and commercialization of these technologies.

One of the most significant challenges in the energy sector is to find sustainable and cost-effective solutions to meet the increasing energy demands of the world's population. The development of sustainable energy technologies is essential to address global challenges of climate change, energy security, and economic growth. To meet these challenges, there is a need for new and innovative materials that can enhance the performance of energy systems. In this case, materials research has contributed significantly to the development of renewable energy technologies. For example, the development of solar cells has been made possible by advances in the understanding of the properties and behavior of semiconductors. The use of new materials, such as perovskites, has led to significant improvements in the efficiency of solar cells (Zhang et al., 2022). Similarly, the development of wind turbines has been made possible by advances in the understanding of materials, such as high-strength alloys, that can withstand the harsh conditions of offshore wind farms (Tong, 2019). In another example, thermoelectric materials can convert heat into electricity and have potential applications in waste heat recovery (Yang et al., 2018).

Energy storage is another critical area where materials science has played a vital role. Particularly, energy storage is essential for renewable energy systems to address the variability of solar and wind energy. The development of materials for energy storage has led to the commercialization of lithium-ion batteries, which are widely used in electric vehicles and consumer electronics (Xu et al., 2023). However, there is a need for new materials that can store more energy (in terms of energy density) and be cost-effective and environmentally sustainable.

Hydrogen is another potential clean energy source that can be used for a variety of applications, from transportation to power generation. Materials science plays a crucial role in the development of materials for hydrogen storage and transportation. Researchers are exploring a range of materials, such as metal hydrides, which can store hydrogen safely and efficiently (Tarasov et al., 2021).

Carbon capture and storage technologies are essential for reducing carbon emissions from industrial processes and power plants. Materials science has contributed to the development of materials that can effectively capture carbon dioxide while being cost-effective and sustainable. Metal-organic frameworks, for example, are highly porous and can selectively capture carbon dioxide from flue gas streams (Pettinari and Tombesi, 2020).

Improving energy efficiency is one of the most effective ways to reduce energy consumption and greenhouse gas emissions. In this case, materials science has contributed significantly to the development of materials for energy-efficient systems. For example, light-emitting diodes (LEDs) are more energy-efficient than traditional lighting systems (Bessho and Shimizu, 2012). Another example is aerogel, which can improve the insulation properties of buildings, reducing energy consumption for heating and cooling (Zhang et al., 2023).

In summary, materials science plays a vital role in the development of sustainable energy technologies. The use of materials has enabled the optimization and commercialization of renewable energy technologies, as seen in solar cells and wind turbines. Materials science has also contributed to the development of materials for energy storage, hydrogen energy, carbon capture, and energy conversion technologies. Furthermore, improving energy efficiency is one of the most effective ways to reduce energy consumption and greenhouse gas emissions. The field has contributed significantly to the development of materials for energy-efficient systems, such as LEDs and aerogels, which have the potential to reduce energy consumption and

FIGURE 1.5 Progress of materials research in LED has enabled the development of efficient LED lighting with various color illumination.

mitigate the effects of climate change. The continued advancement of materials science will be crucial for meeting the energy challenges of the future and transitioning toward a sustainable energy future (Figure 1.5).

1.5 TRENDS IN MATERIALS RESEARCH

Without a doubt, energy is the driving force behind modern civilization. As the world continues to seek sustainable energy sources, the focus has shifted toward researching and developing materials that can improve the efficiency, durability, and performance of energy conversion systems. Some of the trends in materials for energy research that have emerged in recent years are briefly described in this section (Figure 1.6).

Energy storage is an essential part of renewable energy systems. The current challenge is to find materials that can store energy efficiently, while also being cost-effective and environmentally sustainable. One of the most promising materials

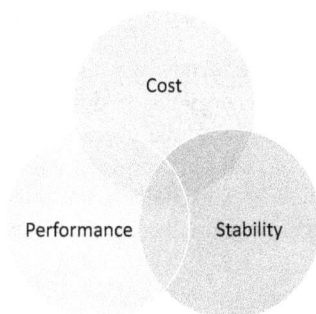

FIGURE 1.6 Trends in materials research for energy systems and devices.

in this field is graphene. Graphene-based batteries can store more energy and charge faster than traditional lithium-ion batteries, making them ideal for use in electric vehicles and other energy storage applications (Olabi et al., 2021). There has been an increasing trend in graphene research in general for the past few years. The application of graphene in energy storage application will see a significant uptake in the coming years.

Solar energy has also become an increasingly popular source of renewable energy. One of the main challenges in this field is to improve the efficiency of solar cells. Silicon-based solar cells have dominated the market for decades, but researchers are now looking at alternative materials, such as perovskites, that have the potential to be more efficient and cheaper. Perovskites are a type of crystal that can be made into thin films and have already shown promising results in laboratory settings (Chen et al., 2023). For scale-up and commercial application, the stability and cost of the solar cell need to be improved.

Another upcoming trend is hydrogen energy. Hydrogen has emerged as a potential clean energy source that can be used for a variety of applications, from transportation to power generation (Rosen and Koohi-Fayegh, 2016). The current challenge in this field is to find materials that can efficiently store and transport hydrogen. New materials that can store hydrogen safely and efficiently are needed. The current metal hydride materials require proper and extensive research for better storage capability (Tarasov et al., 2021).

As a way of promoting sustainable usage, organic materials have been the focus of energy devices. The challenge is to find suitable organic materials that can efficiently convert energy while also being durable and cost-effective. A wide range of organic materials that can be used in flexible and lightweight devices has been explored, especially wearable energy devices' applications (Zhang and Park, 2019).

In conclusion, materials research plays a crucial role in advancing sustainable energy technologies. New materials are explored, from graphene to perovskites to metal oxide frameworks in order to improve the efficiency, durability, and performance of energy systems or devices. With continued research and development, these materials have the potential to transform the way we generate, store, and use energy. Ultimately paving the way for a more sustainable future.

1.6 CONCLUSION

In summary, this chapter has covered crucial aspects of the energy landscape, addressing global challenges, exploring energy conversion principles, and highlighting the pivotal role of materials. Emphasizing the urgency of transitioning to sustainable energy, the emerging trends in materials for energy research have been discussed. This chapter serves as a guide for stakeholders for collaborative efforts to shape a cleaner and more resilient energy future through innovative materials solutions.

REFERENCES

Bejan, A. (2016). *Advanced Engineering Thermodynamics* (Wiley, New York).

Bessho, M., and Shimizu, K. (2012). Latest trends in LED lighting. *Electronics and Communications in Japan* 95, 1–7.

Butler, C., Parkhill, K.A., and Luzecka, P. (2018). Rethinking energy demand governance: Exploring impact beyond 'energy' policy. *Energy Research and Social Science* 36, 70–78.

Chen, Y., Zhang, M., Li, F., and Yang, Z. (2023). Recent progress in perovskite solar cells: Status and future. *Coatings* 13, 644.

IRENA.org. International Renewable Energy Agency (IRENA). (2023). Renewables Competitiveness Accelerates, Despite Cost Inflation. Available at https://www.irena.org/News/pressreleases/2023/Aug/Renewables-Competitiveness-Accelerates-Despite-Cost-Inflation (accessed November 2023).

Jenkins, K. (2018). Energy poverty and vulnerability: A global perspective. *Energy* 147, 388–389.

Kabeyi, M.J.B., and Olanrewaju, O.A. (2022). Sustainable energy transition for renewable and low carbon grid electricity generation and supply. *Frontiers in Energy Research* 9.

Khan, I., Zakari, A., Ahmad, M., Irfan, M., and Hou, F. (2022). Linking energy transitions, energy consumption, and environmental sustainability in OECD countries. *Gondwana Research* 103, 445–457.

Lostaglio, M., Jennings, D., and Rudolph, T. (2014). Thermodynamic laws beyond free energy relations. 11, arXiv preprint arXiv:1405.2188.

Nevskaya, M.A., Raikhlin, S.M., Vinogradova, V.V., Belyaev, V.V., and Khaikin, M.M. (2023). A study of factors affecting national energy efficiency. *Energies* 16, 5170.

Olabi, A.G., Abdelkareem, M.A., Wilberforce, T., and Sayed, E.T. (2021). Application of graphene in energy storage device: A review. *Renewable and Sustainable Energy Reviews* 135, 110026.

Pettinari, C., and Tombesi, A. (2020). Metal-organic frameworks for carbon dioxide capture. *MRS Energy and Sustainability* 7, 35.

Rosen, M.A., and Koohi-Fayegh, S. (2016). The prospects for hydrogen as an energy carrier: an overview of hydrogen energy and hydrogen energy systems. *Energy, Ecology and Environment* 1, 10–29.

Scovronick, N., Budolfson, M., Dennig, F., Errickson, F., Fleurbaey, M., Peng, W., Socolow, R.H., Spears, D., and Wagner, F. (2019). The impact of human health co-benefits on evaluations of global climate policy. *Nature Communications* 10, 2095.

Steinbacher, K., and Röhrkasten, S. (2019). An outlook on Germany's international energy transition policy in the years to come: Solid foundations and new challenges. *Energy Research and Social Science* 49, 204–208.

Stevens, P. (2012). The 'shale gas revolution': Developments and changes. Chatham House Briefing Paper 12.

Tarasov, B.P., Fursikov, P.V., Volodin, A.A., Bocharnikov, M.S., Shimkus, Y.Y., Kashin, A.M., Yartys, V.A., Chidziva, S., Pasupathi, S., and Lototskyy, M.V. (2021). Metal hydride hydrogen storage and compression systems for energy storage technologies. *International Journal of Hydrogen Energy* 46, 13647–13657.

Tong, C. (2019). Advanced materials enable renewable wind energy capture and generation. In: *Introduction to Materials for Advanced Energy Systems* (Springer, New York), pp. 379–444.

Wang, D., Liu, X., Yang, X., Zhang, Z., Wen, X., and Zhao, Y. (2021). China's energy transition policy expectation and its CO_2 emission reduction effect assessment. *Frontiers in Energy Research* 8.

Xu, J., Cai, X., Cai, S., Shao, Y., Hu, C., Lu, S., and Ding, S. (2023). High-energy lithium-ion batteries: recent progress and a promising future in applications. *Energy and Environmental Materials* 6, e12450.

Yang, L., Chen, Z.G., Dargusch, M.S., and Zou, J. (2018). High performance thermoelectric materials: progress and their applications. *Advanced Energy Materials* 8, 1701797.

Zhang, P., Li, M., and Chen, W.C. (2022). A perspective on perovskite solar cells: emergence, progress, and commercialization. *Frontiers in Chemistry* 10.

Zhang, X., Hu, H., Hou, Q., and Liu, W. (2023). Research progress of cellulose-based aerogel in heat preservation and insulation. *Chung-Kuo Tsao Chih/China Pulp and Paper* 42, 86–93.

Zhang, Y., and Park, S.J. (2019). Flexible organic thermoelectric materials and devices for wearable green energy harvesting. *Polymers* 11, 909.

2 Photovoltaic Fundamental of Energy Conversion

Hieng Kiat Jun and Foo Wah Low

2.1 INTRODUCTION AND BRIEF HISTORY

A photovoltaic (PV) system, also known as a solar electric system, is a technology that converts sunlight into electricity. It is composed of solar panels that contain PV cells, which convert the energy from the sun into direct current (DC) electricity. This DC electricity can then be converted into alternating current (AC) electricity using an inverter, which allows it to be used in homes and businesses. PV systems can be used for a variety of applications, such as powering homes, businesses, and even vehicles. They are becoming increasingly popular as a source of renewable energy. This is because PV systems do not produce emissions and are considered a sustainable way to generate electricity.

A solar cell and a PV cell are essentially the same thing. A solar cell is a device that converts sunlight into electricity through the PV effect. The main difference between the two is that solar cell is a generic term that refers to any device that converts sunlight into electricity, while PV cell specifically refers to a type of solar cell that uses the PV effect to generate electricity. Solar cells can be made using a variety of materials, such as silicon (Si), cadmium telluride (CdTe), or copper indium gallium selenide.

The history of PV can be traced back to the 19th century, when scientists first discovered the PV effect. In 1839, French physicist Alexandre-Edmond Becquerel observed that certain materials would produce small amounts of electrical current when exposed to light (Palz, 2010). This phenomenon is known as the PV effect, and it forms the basis of modern PV technology.

In the early 20th century, researchers began experimenting with using the PV effect to generate electricity. In 1941, Russell Ohl discovered that silicon, a common semiconductor material, could be used to create a PV cell (Riordan and Hoddeson, 1997). However, the efficiency of these early PV cells was low and they were not cost-effective for widespread use at that time.

In the 1950s, Bell Labs developed the first practical silicon PV cell, which had an efficiency of 6%. This marked the beginning of the modern PV industry (Blandford and Watkins, 2009). Over the next few decades, researchers continued to improve the efficiency of PV cells and lower their cost. In the 1970s, the oil crisis led to increased interest in renewable energy sources, and PV technology began to be used

DOI: 10.1201/9781003314424-2

in a variety of applications. The increased interest was mainly attributed by oil companies venturing into solar firms (Wayback Machine, 2012).

By the early 21st century, PV technology had advanced significantly. The efficiency of PV cells has increased to over 20%, and the cost of PV systems has dropped significantly (Jaeger, 2021). As a result, PV systems began to be used for large-scale power generation and powering homes and businesses. Today, PV systems are becoming increasingly popular as a source of renewable energy. They are used in a variety of applications, such as powering homes, businesses, and even vehicles. They are also being integrated into the electricity grid as a source of distributed energy generation. In recent years, PV panel production has grown exponentially, with some countries like China and the US leading the way in terms of production and installation (IEA, 2015). With the cost of the solar panel dropping and the technology improving, the future of PV looks promising.

2.2 BASIC CHARACTERISTICS AND CHARACTERIZATION OF PHOTOVOLTAICS

The main characteristic of PV is the energy conversion capability. Specifically, PV cells convert light into electricity through the PV effect. In most cases, DC electricity is generated, which can then be converted into AC using an inverter. The performance of a PV cell is expressed in efficiency. The efficiency of a PV cell is a measure of how effectively it converts light into electricity. Typical commercial silicon-based PV cells have an efficiency of around 20% (Schultz et al., 2007).

Most commercial PV cells are durable and have a long lifespan. They are designed to withstand extreme weather conditions and have minimal maintenance requirements. Aside from that, PV systems are categorized as a clean and renewable source of energy. They do not produce emissions and have a low environmental impact.

One of the salient features of PV systems is their flexibility where they can be installed on a variety of surfaces, with most installations taking place on rooftops. Not only that, the systems can be scaled to meet the energy needs of a variety of applications, from powering a small device or single home to a large-scale power plant. In terms of cost, the cost of PV systems has dropped significantly in recent years, making them increasingly cost-effective compared to traditional sources of electricity derived from fossil fuels (Martin, 2016).

There are several methods used to characterize and evaluate the performance of PV cells and systems. The most frequently used method is current–voltage (I–V) measurement. In this method, the current and voltage responses of a PV cell or system are measured under standard conditions. The standard test condition entails a solar irradiance value of 1,000 W/m² with an exposure period of up to 2.74 hours a day. I–V characteristic can also be measured under varying light intensity and temperature. Such data can provide comprehensive data on the PV cell which is beneficial in designing integrated PV system. The output of the measurement is represented as I–V curve of J–V curve (J represents current density), which is used to determine the efficiency and maximum power point of a PV cell or system (Figure 2.1). From the same I–V or J–V curve, other parameters can be derived,

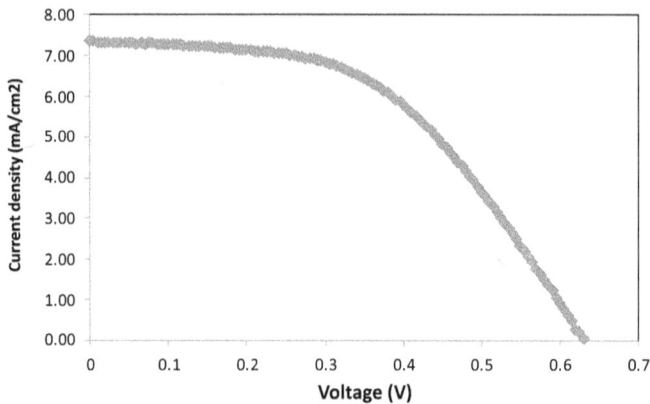

FIGURE 2.1 *J–V* curve of dye-sensitized solar cell using black dye as a sensitizer.

such as fill factor and maximum power point. Fill factor is a measure of quality of a solar cell while the maximum power point is the point where the product of *I* and *V* is the maximum.

In general, most of the PV cells do not respond to the entire spectrum of UV-visible regions. As such, it will be prudent to assess the spectral response of the PV active materials (i.e. materials responsible for the energy conversion). In this method, the electrical output of a PV cell or system illuminated or exposed to different wavelengths of light is measured. The output of the measurement can then be used to determine the cell's spectral response and identify the best wavelength for the highest efficiency.

Another less-discussed characterization of PV is temperature coefficient measurement. In this method, the performance of a PV cell or system is measured as a function of temperature. With increasing temperature, the efficiency of a PV cell usually will decrease. The data can be used to determine the temperature coefficient and identify the operating temperature range of the PV cell or system.

Other characterization or measurement methods include light-induced degradation measurement, current-matching measurement, electroluminescence imaging, and quantum efficiency (QE) measurement. The QE measurement is sometimes discussed in scientific journals pertaining to the PV materials' characteristics. QE denotes the proportion of photons that transform into electric current (i.e., gathered carriers) when the cell operates in short circuit conditions. In this method, the number of photocurrent generated by the PV cells per number of incident photons is determined. The data is then used to determine the internal QE of the cells.

2.3 PHOTOCURRENT GENERATION AND PHOTOVOLTAGE

The PV effect is the mechanism by which PV cells generate electricity. This effect occurs when photons of light are absorbed by a material, causing the release of electrons and the creation of an electrical current. In PV cells, the material that absorbs photons is typically a semiconductor, such as silicon. Semiconductors have a specific bandgap energy, which determines the energy required for an electron to be excited from the valence band to the conduction band. When a photon with energy greater than the bandgap energy is absorbed by the semiconductor, an electron is excited from the valence band to the conduction band, leaving a hole behind. These excited electrons and holes are referred to as free carriers, and they are able to move through the semiconductor. A PV cell is designed to collect these free carriers and separate them, creating an electrical current.

The positive and negative terminals of the PV cell are typically made of different semiconductor materials, known as p-type and n-type semiconductors. The p-type semiconductor has an excess of holes, and the n-type semiconductor has an excess of electrons. When the p-type and n-type semiconductors are brought into contact, a depletion region is formed at the interface between them. The depletion region is a region where the free electrons and holes are forced to recombine, creating an electric field. This electric field acts as a barrier, separating the free electrons and holes and creating a voltage across the terminals of the PV cell (Mishra, 2008).

In the $P–N$ junction diode, four current components are present in equilibrium condition, namely, electron drift, electron diffusion, hole drift, and hole diffusion. In equilibrium condition, the net current is zero, which requires the drift and diffusion current of carriers to be equal and opposite. When PV is illuminating under light, it results large drift current with minority carriers (Figure 2.2), which flow from the n-type side to the p-type side. Since current flow is caused by the light, it is known as light-generated current, I_L. The generated photovoltage due to the light, biases the $P–N$ junction in forward bias mode. Generated photovoltage reduces the junction's potential energy barrier and causes diffusion current flowing in the opposite direction to the I_L. But the magnitude of I_L is larger than the forward-biased diffusion

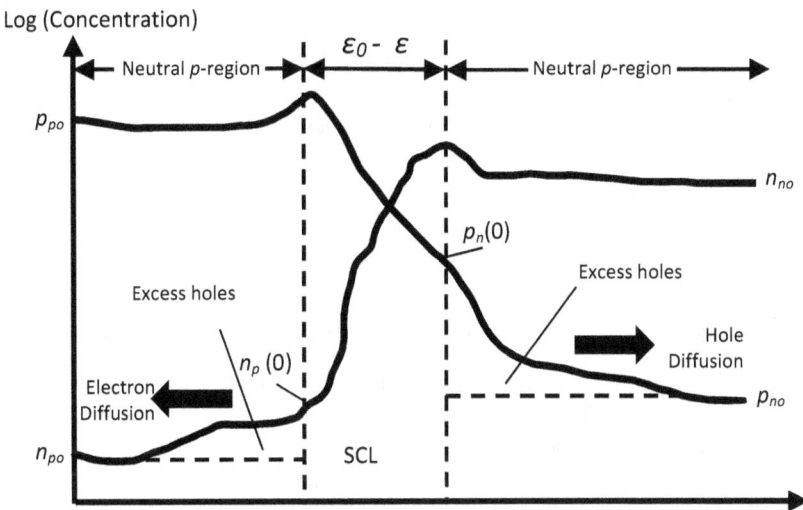

FIGURE 2.2 Minority and majority carriers profile for $P–N$ junction.

current, and hence net current flows from the n-type side to the p-type side. Thus, when light shines on the PV cell, current flow is in the opposite direction respective to the generated voltage.

This voltage is known as the photovoltage, and the current generated by the flow of free carriers is known as the photocurrent. The combination of the photovoltage and photocurrent generates an electrical power output from the PV cell, which can be used to drive an electrical load. The photovoltage and photocurrent vary with the light intensity, temperature, and other environmental factors. Under uniform illumination condition, the generation of charge carrier will occur in the space charge region as well as the quasi-neutral region. The generated carriers in the space charge region will immediately swap due to the electric field. During efficient flow of electric field, the occurrence of recombination centers is quite less. Somehow, there are random motions of generated electron–hole pairs in quasi-neutral regions. Random motion of generated carriers eventually will be closer to the edge of the space charge region due to the force of the electric field. Only minority carriers will cross the junction as carriers always head down due to the energy force. The swapping of the *P–N* junction will induce a potential difference under light illumination. This phenomenon of photovoltage generation is known as PV effect.

There are seldom cases of random motion of carriers generated in the quasi-neutral region crossing the junction. In the event of this occurrence for these minority carriers, they will travel an average distance, L_n or L_p, before recombining with the opposite type of carrier. L_n and L_p are denoted as the diffusion length of electrons and holes, respectively. The minority carriers in the quasi-neutral region will be stuck at the edge of the *P–N* junction and will not contribute to the photovoltage. In rare cases, once the minority carriers manage to cross into the junction, the generated carriers within the width will be $(L_n + W + L_p)$. The rate of minority carriers depends on the intensity of light. In fact, the generation of carriers for a PV cell results in a significant increase in minority carrier concentration, while the majority carriers remain unaffected (refer to Figure 2.2). Only if the minority carriers within the diffusion length are able to cross the junction. Due to the rapid drifting of charge carriers in the *P–N* junction caused by the electric field, the minority carrier concentration at the edge of the junction is nearly equal to the equilibrium concentration. In another way, for those minority carriers drifting far away from the junction, there is a low chance of boarding toward the junction edge; hence, the minority carriers' concentration increases in the quasi-neutral region. The carrier concentration attains a steady state against recombination in the quasi-neutral region when it exceeds the diffusion length.

2.4 CHARGE SEPARATION, RECOMBINATION, AND EFFICIENCY

In a solar cell, charge separation and recombination refer to the processes by which the electrons and holes (positive charge carriers) generated by the absorption of photons are separated and subsequently recombined. When a photon is absorbed by a solar cell, it excites an electron in the valence band of the semiconductor material to the conduction band, leaving a hole in the valence band. These excited electrons and holes are known as free carriers, and they are able to move through the semiconductor material. The goal of a solar cell is to separate these free carriers and create an electrical current.

Charge separation is achieved by creating a built-in electric field at the junction between the p-type and n-type semiconductor materials. The p-type semiconductor has a surplus of holes, while the n-type semiconductor has a surplus of electrons. When the p-type and n-type semiconductors are brought into contact, a depletion region is formed at the interface between them. This depletion region acts as a barrier, separating the free electrons and holes and creating a voltage across the terminals of the solar cell (refer Figure 2.3).

Once the electrons and holes are separated, they flow in opposite directions to the terminals of the solar cell, generating an electrical current. However, this separation is not perfect and some of the free carriers will recombine before they reach the terminals of the cell. This recombination results in a loss of the electrical current and it reduces the solar cell efficiency.

Recombination can occur at various defects in the solar cell material such as grain boundaries, recombination centers, and surface states. When a working PV cell is under illumination, electrons and holes will be consequently generated, thus it will influence the equilibrium conditions of material. The process of recombination of charge carriers is in the opposite condition to the generation of charge carriers. In this comprehensive process, excited electrons will fall back from the conduction band into the valence band, reoccupying an empty energy state in the valence band. Henceforth, the electron–hole pairs are destructed. In other words, the recombination process can be considered as a restoring process of the carrier's mechanism. The recombination issue is not desirable in PV operation because it will be limiting the PV efficiency performance.

Assuming the concentration of electrons and holes in equilibrium condition in n_0 and p_0, when the light falls on the PV device, non-equilibrium concentration of electrons and holes will be induced with n and p. Therefore, the excess charge carrier concentration of electrons and holes under illumination condition, Δp and Δp, respectively, can be written as follows:

$$\Delta n = n - n_0 \tag{2.1}$$

$$\Delta p = p - p_0 \tag{2.2}$$

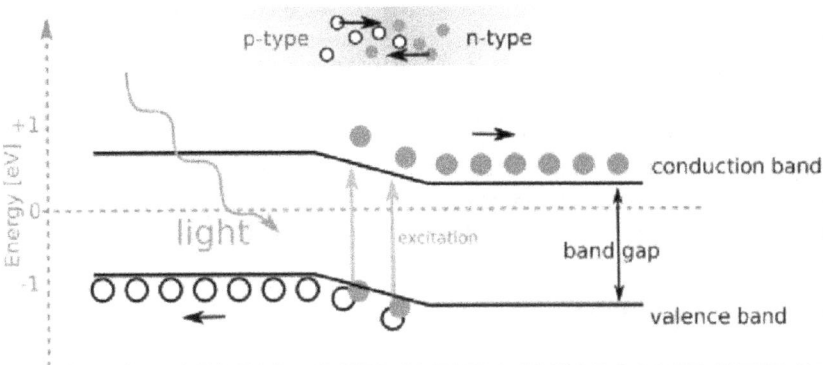

FIGURE 2.3 Electron–hole pairs generation under illuminated light.

Charge carrier lifetime is defined that the period for a charge carrier returned to the ground state from the excited state during the occurrence of recombination. Indeed, it should consider in average time taken for charge carriers to travel to the ground state, the symbol is tau, τ. τ_n and τ_p denoted for electrons and holes with time unit, respectively. The τ value could be as low as nanoseconds to milliseconds. The charge carrier lifetime depends on the excess carrier concentration and rate of recombination, R, which is given as follows:

$$\tau_n = \frac{\Delta n}{R} \tag{2.3}$$

$$\tau_p = \frac{\Delta p}{R} \tag{2.4}$$

In short, low structural defects and low undesirable impurity concentration are deserved to obtain a high carrier lifetime.

The efficiency of solar cell is affected by the recombination rate. The higher the recombination rate, the lower the efficiency of the solar cell. Therefore, one of the main goals of solar cell research is to reduce recombination and increase charge separation to improve solar cell efficiency. This can be achieved by using high-quality semiconductor materials, reducing defects in the solar cell material, and optimizing the design of the solar cell. To increase solar cell efficiency, researchers have developed various techniques such as passivating the surfaces of the solar cell, using doping techniques and multi-junction cells to increase the amount of energy that can be harvested from the sun.

2.5 CONCLUSION

This chapter covers the historical development, basic characteristics, and essential processes of photovoltaics. It began by tracing the evolution of PV technology, highlighting its pivotal role in solar energy utilization. The exploration of fundamental PV characteristics, including photocurrent generation and photovoltage, provides insights into the intricate interplay of photons, semiconductors, and electrical circuits. The discussion extends to crucial phenomena such as charge separation and recombination within PV cells. Understanding these processes is key to optimizing device performance. Efficiency emerged as a central theme, emphasizing the need for a holistic approach that considers material properties, device design, and fabrication processes. As the demand for sustainable energy solutions rises, this chapter lays the groundwork for future advancements in PV technology, promoting the development of more efficient and widely adopted solar energy systems.

ACKNOWLEDGMENT

HK Jun thanks the Air Force Office of Scientific Research for the research support under award number FA2386-21-1–4106.

REFERENCES

Blandford, R., and Watkins, M. (2009). Bell labs demonstrates the first practical silicon solar cell. *APS News* 18, 2.

IEA (2015). Snapshot of global PV 1992–2014. In: *Photovoltaic Power Systems Programme*, IEA, pp. 5–6.

Jaeger, J. (2021). *Explaining the Exponential Growth of Renewable Energy*. World Resources Institute, Washington, DC, pp. 1–9.

Martin, C. (2016). *Solar Panels Now So Cheap Manufacturers Probably Selling at Loss*. Bloomberg, London.

Mishra, U. (2008). *Semiconductor Device Physics and Design*. Springer, New York.

Palz, W. (2010). *Power for the World: The Emergence of Electricity from the Sun*. Pan Stanford Publishing, Belgium, p. 6.

Riordan, M., and Hoddeson, L. (1997). Origins of the pn junction. *IEEE Spectrum*, 34, 46–51.

Schultz, O., Mette, A., Preu, R., and Gluntz, S.W. (2007). Silicon solar cells with screen-printed front side metallization exceeding 19% efficiency. In: *Proceedings of the 22nd European Photovoltaic Solar Energy Conference and Exhibition*, Milano, pp. 6–9.

Wayback Machine. (2012). *The 1970s Energy Crisis Archived*, February 15, 2012. Available at https://web.archive.org/web/20120215190032/https://cr.middlebury.edu/es/altenergylife/70%27s.htm (accessed October 2023).

3 Photovoltaic
Graphene Materials and Their Effects in Perovskite Solar Cells

*Foo Wah Low, Muhammad Kashif,
Mohammad Aminul Islam, Hieng Kiat Jun,
Cheen Sean Oon, and Chun Hong Voon*

3.1 INTRODUCTION

In the 21st century, rapid technological development and the busy lifestyle of humans, often reliant on digitalization, have significantly increased the demand for electricity. Undeniably, fossil fuels remain our primary source for generating the essential energy needed for life. However, this reliance contributes to climate change through increased combustion, ultimately leading to incidents such as the devastating bushfires in Australia. Henceforth, the promotion of renewable energy as an alternative electricity source, such as solar energy, has been introduced for the betterment of human society. Compared to other renewable energy sources, solar energy stands out as the most reliable option, capable of converting sunlight into useful electrical energy—offering a clean and sustainable solution.

One solution is to transition our daily needs to a smarter lifestyle by incorporating technologies such as Artificial Intelligence with the Internet of Things, smart home appliances, and integrating solar PV or fuel cells into vehicles. However, these advancements are often considered "high-cost" compared to existing analog products. In the case of low-power products, there are still alternative ways, such as generating energy from solar cells. Solar energy is a redundant and renewable source that can be easily harnessed from the sun.

Currently, solid-state-based Perovskites (PSCs) PV technology has become a popular subject in the PV industry due to its high-efficiency performance and good light absorption characteristics. Despite the potential advantages, this technology is still in the preliminary stage and is not yet available in the market compared to other PV technologies. The most common electrode layer currently used is gold, which increases the overall cost of this technology and places a burden on customers. Moreover, PSCs are susceptible to rapid deterioration in moist environments, such as during rainy days, and the decay products may corrode the metal electrodes, leading to a reduction in the performance of the PSCs. To protect the solar cell, heavy encapsulation is added, contributing to increased production costs and weight. Considering

DOI: 10.1201/9781003314424-3

the interface and electron transporting layer (ETL), PSCs exhibit various carrier generation, electron transportation, and recombination blocking behaviors. The transfer rate of electrons at the ETL interface is expected to be slower than that at the hole transporting layer (HTL) in PSCs. This condition influences the hysteresis behavior in efficiency measurements and also affects the stability of PSCs, such as the migration of ions.

The photovoltaic (PV) effect was discovered by Becquerel, a French physicist, in 1839. Later, in 1876, British scientists Adams and other researchers found that semiconductors of selenium could generate electricity under sunlight (Blume, n.d.). In 1883, Fritts successfully invented a new semiconductor junction solar cell composed of a thin gold layer with germanium. However, the capability of his innovation was only about 1% (Fritts, 1883). A significant milestone in the PV industry occurred in 1954 when Pearson and other researchers from the US Bell Labs introduced the primary Crystalline Silicon Solar Cell (CSSC), achieving a power conversion efficiency (PCE) of up to 4.5% (Chapin et al., 1954). Nevertheless, these early-generation solar cells were highly costly and environmentally unfriendly.

Among PV generations, third-generation solar cells have been recently invented, with dye-sensitized solar cells (DSSCs) being the most popular. However, these PV cells have two major disadvantages in sunlight absorption. First, the manufacturing of DSSC involves a thickness that makes it challenging to absorb light using a thin absorption layer in solid-state solar cells. Second, light bleaching occurs in DSSC applications, making them less preferable for everyday use. To address these issues, Japanese scientists, led by Kojima and his team, discovered that organic metal halides have similar properties for absorbing sunlight, eliminating the need for dye functionalization (Kojima et al., 2009). The pioneering perovskite solar cells (PSCs), which operate entirely in a solid-state situation, achieved a PCE of 9.7%, as announced by Kim and his team in 2012 (Kim et al., 2016). PSCs have garnered global attention and have risen rapidly in popularity due to their lower cost compared to other PV generations and superior performance. The highest PCE of PSCs was successfully achieved in 2016, reaching up to 22.1%.

3.2 CRYSTALLINE SI-BASED PHOTOVOLTAIC

Silicon is the most commonly used material for PV applications. Silicon-based solar cells are capable of capturing photons from sunlight with an energy greater than 1.1 eV. However, CSSCs do not cover the full range of energy in the entire solar spectrum. The primary factor limiting the performance of silicon-based solar cells is the fixed bandgap of silicon material itself. Moreover, the fabrication of tandem solar cells requires materials with different bandgaps for each layer. Therefore, multijunction silicon-based solar cells show promise with the modification of silicon material by reducing the size of silicon crystals. The optical and electrical properties of semiconductor materials undergo changes when the crystalline size becomes smaller than the Bohr exciton radius (i.e., the Bohr exciton radius of silicon is <5 nm). Consequently, silicon is decorated with other materials exhibiting different properties. For example, silicon materials can be incorporated with high-bandgap metal oxides such as SiO_2 and Si_3N_4. The arrangement of silicon with high-bandgap metal

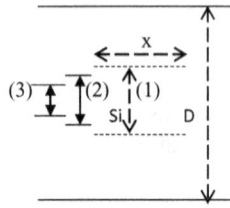

FIGURE 3.1 Schematic bandgap diagram of crystalline silicon in a dielectric matrix.

oxide materials tends to alter the respective bandgap. The energy bandgap diagram of silicon embedded with metal oxide material is illustrated in Figure 3.1. In this arrangement, the changes in electrical and optical properties are known as quantum confinement carriers (labeled "1"). As the size of silicon nanoparticles decreases, a lower Bohr exciton radius is required. In other words, the bandgap narrows between the lowest energy state in the conduction band and the higher occupied energy state in the valence band (labeled "2"). The bandgap becomes even narrower as the nanoparticle size decreases, promising a higher effective bandgap (labeled "3"). The equation describing the relationship between bandgap (E_g in eV) and nanocrystal size is summarized in Equation 3.1 (Conibeer et al., 2006).

$$E_g = E_g + \frac{k}{d^2} \tag{3.1}$$

where k = quantum confinement parameter, d = diameter of nanoparticles (nm), and E_g = bandgap of material.

The increased bandgap, influenced by small silicon nanocrystals and ultimately resulting in quantum confinement, is referred to as quantum dots (QDs). The narrower spacing of silicon QD bonds leads to a wider bandgap when incorporated with other semiconductor materials. By tuning the size of Si QD, the optimized bandgap is approximately 2.7 eV. In recent years, there has been a significant amount of research on silicon nanocrystals decorated with metal oxide due to their potential applications in PV.

3.3 III–V SEMICONDUCTORS MULTIJUNCTION-BASED PHOTOVOLTAIC

A III–V semiconductor is a compound, such as gallium nitride (GaN) or gallium arsenide (GaAs). Each of these compounds has three (III) and five (V) valence electrons, providing high electron mobility. Consequently, GaN or GaAs semiconductors are widely used in PV applications. GaAs, in particular, is notable for being the first multijunction PV cell that combines with single-junction solar cell technologies. In the context of multijunction PV structures, cells are stacked, resulting in a lower energy bandgap.

GaAs can absorb high-energy photons, but it can also transmit photons in the circuit loop with lower energies. The Shockley–Queisser limit refers to the maximum efficiency applicable for a single-material PV. Assuming that each energy gap gives

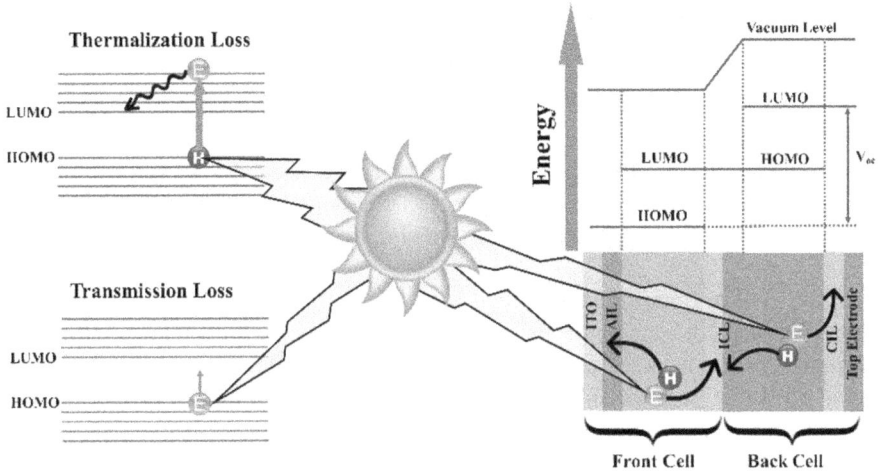

FIGURE 3.2 Schematic diagram of an electronic energy level of thermalization and transmission losses.

rise to only one electron–hole pair, the theoretical maximum PCE can be achieved up to 31%. In reality, there is a total loss of photon energy, approximately 56%, mainly attributed to thermalization and transmission losses (refer Figure 3.2).

Multijunction GaAs PV cells have achieved performances of up to 47.1% with a triple-junction cell (France et al., 2022). While high-efficiency GaAs PV cells are considered expensive, they are 40% more efficient than other types of PV.

3.4   THIN-FILM PHOTOVOLTAIC

Silicon-based solar technologies are generally categorized under first-generation technologies, while thin film–based PV is considered second-generation technology. The primary development objectives of second-generation thin film–based PV include reducing the fabrication cost of silicon wafer-based solar PV modules and achieving higher efficiency than silicon-based PV technologies. Thin-film PV technology involves the use of very thin layers of semiconductor material to convert sunlight into electricity. Materials commonly used in thin-film PV include amorphous silicon, cadmium telluride (CdTe), copper indium gallium selenide, and thin-film CSSC technologies. Another definition of thin film refers to the "random nucleation and growth process of individually condensing/reacting atomic/ionic/molecular species on a substrate." The structural, chemical, and physical properties of the specific film depend on the deposition parameters. In fact, the thickness of the film is a crucial parameter, ranging from a few nanometers to several tens of micrometers. Thin materials obtained through processes such as wafer thinning or depositing clusters of microscopic species, as in screen printing, are referred to as thick films.

In principle, any semiconductor material can be used for thin-film fabrication. However, only a few materials can achieve reasonable efficiency for light-to-electricity conversion. Ideally, the selection of semiconductor materials for thin-film fabrication must fulfill certain requirements to achieve high-efficiency conversion. These requirements are as follows:

- Bandgap between 1 and 1.5 eV.
- Absorption coefficient of the material should be high, 10^4–$10^6 cm^{-1}$.
- Recombination rate of generated charge carriers should be low.
- Diffusion length of generated charge carriers should be high.
- Materials should be widely available, reproducible, and non-toxic.

Basically, the operation of a solar cell is based on the generation of photo-voltage upon illumination. Semiconductor material is used to form a junction to facilitate the separation of charge carriers. There are two types of junction formation: homo junction (two similar materials) and hetero junction (two different materials). Thin-film technologies mostly apply inorganic materials for fabrication.

3.5 NANOCOMPOSITE SOLAR CELLS

Nanocomposite solar cells represent an innovative frontier in PV technology, leveraging the unique properties of nanomaterials to enhance solar energy conversion efficiency. By incorporating nanoscale structures or composite materials into the design, these solar cells aim to overcome traditional limitations, such as suboptimal light absorption or charge carrier mobility. The synergistic effects of combining different materials at the nanoscale allow for precise control over the solar cell's properties, leading to improved performance and promising advancements in the quest for more efficient and sustainable solar energy solutions. Examples of nanocomposite materials used in solar cells include perovskite and dye sensitizers.

3.5.1 PEROVSKITES SOLAR CELLS (PSCs)

PSCs are solid-state PV cells that comprise several materials, with the main component being a perovskite. Perovskite can refer to any material that forms a crystal structure similar to calcium titanium oxide. Consequently, PSCs exhibit higher efficiency in converting photons from sunlight into electricity. This superior efficiency in terms of PCE positions PSCs ahead of other solar energy technologies. Recently, PSCs have emerged as a prominent sector of research in addition to other PV technologies. Figure 3.3 shows the schematic layout for the PSCs structure.

There are two categories of PV technologies in the past: wafer-based PV and thin-film cell PV. Then, PSCs are categorized as emerging thin-film cell PV and are classified under the third generation in the PV industry. There are some other types of solar cells produced with PSCs, such as organic-based PV, quantum Dot PV, and DSSCs. Recent research on PSCs has shown that the efficiency gradually increased by 18.3% from 2009 until 2016 (Park et al., 2016). There are many advantages to PSCs compared to other PV technologies. The most attractive and essential advantage is their achievable highly efficient performances compared to other solar technologies. Remarkably, there is an increase from 3.8% to 22% from 2009 to 2016. Scientists and researchers believe that the PCE can be enhanced further as more

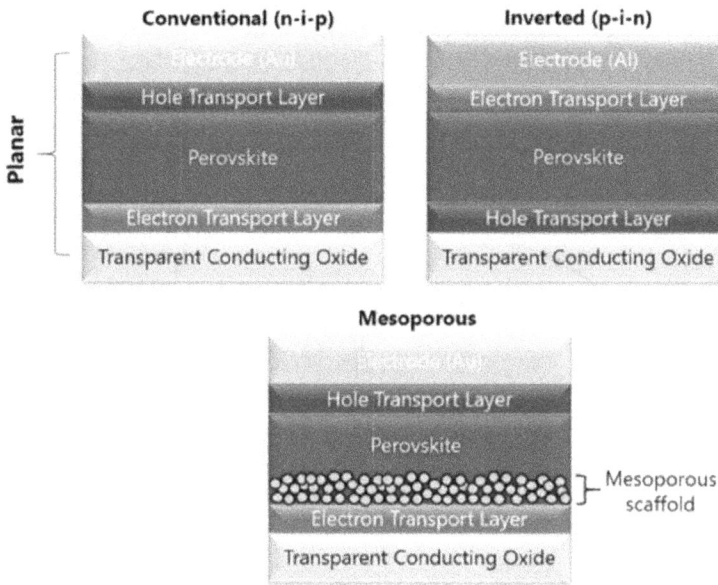

FIGURE 3.3   Basic PSCs structure (Valsalakumar et al., 2023).

studies focus on PSCs. In fact, PSCs have to be tolerant to defects, have a long carrier lifetime, long diffusion distance of photo-generated carriers, and a high absorption coefficient to obtain better PCE performance.

In fact, the cost of PSCs is also relatively low compared to other PV cells, as shown in Figure 3.4. This is because PSCs are made with man-made materials, unlike traditional solar cells that are made with crystalline silicon. Indeed, silicon material is extracted from the Earth and needs further processing before being made into a highly efficient solar cell, which makes traditional solar cells expensive. Unlike other traditional solar cells, PSCs are simply produced by a process called "solution processing," which follows the same procedures used for printing newspapers.

Furthermore, PSCs share many outstanding properties with other thin-film layer solar cells. Importantly, PSCs are more flexible and lightweight compared to traditional silicon solar cells. The special lightweight property of PSCs facilitates easier installation on roofs or walls, requiring less physical stress. Moreover, PSCs are also semi-transparent, aiding in more effective sunlight absorption. In the same vein, PSCs are easier to manufacture and excel in gathering sunlight. These properties make them more efficient and popular for consumers. Additionally, there are further advantages of PSCs in the future, and they represent a significant potential benefit for our daily lives, especially concerning the environmental impact related to the lead-based perovskite absorber. Researchers are studying alternative substitutions for lead, given its high toxicity to the environment. This effort is crucial for the future, aiming to reduce and eliminate the toxicity of lead to preserve our polluted environment for future generations. As such, PSCs have the potential to adopt a lead-free structure, fulfilling the basic aim of using solar cells to generate electricity.

Estimated cost/watt

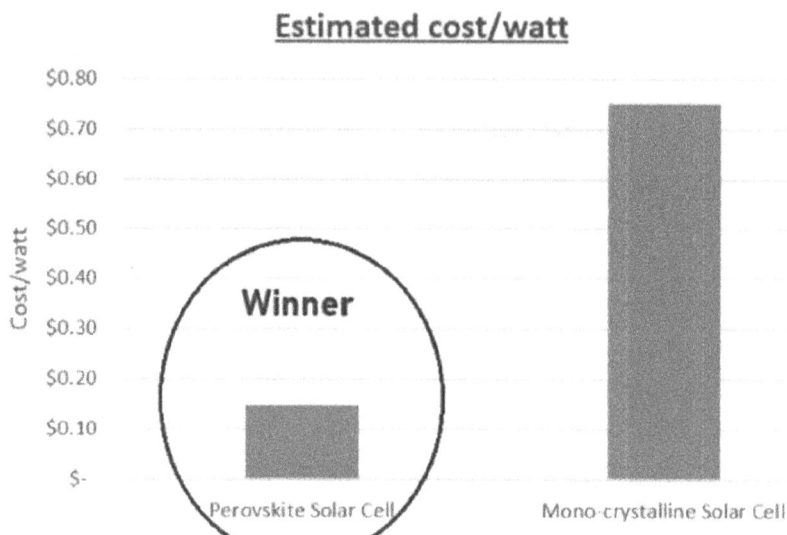

FIGURE 3.4   Cost of PSCs compared to silicon-based solar cell (Qian et al., 2019).

3.5.2   STRUCTURE OF PSCs

Generally, perovskite is defined as $CaTiO_3$, with the molecular structure stated in the *ABX₃* formula. Perovskite has attracted much attention due to its layered structures, characterized by a cubic lattice-nested octahedral shape, as well as its optical, unique thermal, and electrochemical properties. The perovskite used in solar cells is typically a compound of an organic metal halide and inorganic metal halide, such as $CH_3NH_3^+$, MA^+, or $CH(NH_2)_2^+$. FA^+ is positioned around the center of the cubic lattice, while the metal halide with Group 2 metals as cations, such as lead or tin, acts as the core of the octahedra. Meanwhile, halogens, such as chlorine or bromine, act as the apex of the octahedra. Figure 3.5 illustrates the structure of perovskite in solar cells.

3.5.3   WORKING PRINCIPLE OF PSCs

The structure of PSCs typically provides four features for the solar cell (refer Figure 3.6). First, the components maintain excellent photoelectric properties, with high optical absorption coefficients that can reach up to 10^4 cm^{-1}, and lower exciton binding energy. Second, solar cells that use perovskite as a layer to absorb light can increase efficiency in gathering sunlight. Besides that, electrons, holes, and dielectric constant can be obtained by the materials in a more effective way. Last, holes and electrons can be transported together, and the distance for transmission is approximately 100 nm and may go up to 1 μm.

   These features contribute to the high open-circuit voltage (V_{OC}) and the short-circuit current density (J_{SC}). Initially, the perovskite layer absorbs photons, producing excitons (electron–hole pairs) when exposed to sunlight. Subsequently, free charge carriers can be formed by these electron–hole pairs, generating current due to the difference in

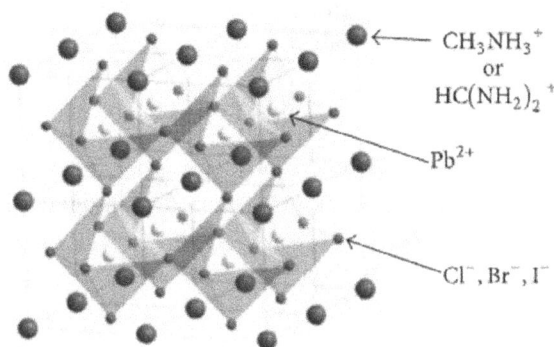

FIGURE 3.5 Structure of PSCs (Chen, 2023).

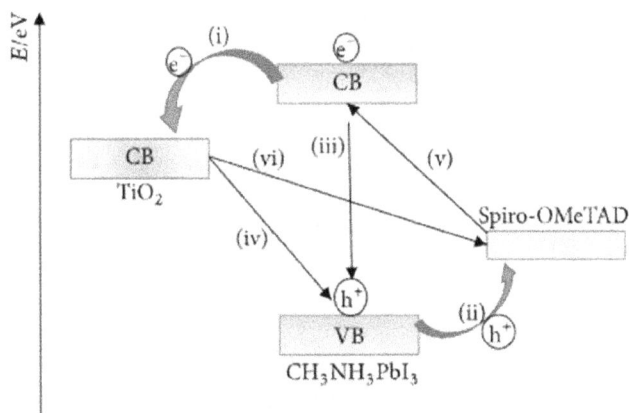

FIGURE 3.6 Mechanism of PSCs.

exciton energy levels for the perovskite material. The distance for diffusion and the duration of the carrier are long, as the probability of carrier recombination for $CH_3NH_3PbI_3$ (MAPbI$_3$) is low, coupled with the high mobility of the carrier. The lifetime and the distance of the carrier are directly proportional to the efficiency of the PSCs.

Next, the free holes and electrons are collected by the hole transport material (HTM) and electron transport material (ETM), respectively. Then, electrons are transferred from perovskite to the ETL. Afterward, the electrons are collected by fluorine doped tin oxide (FTO) from the ETL. Besides that, holes are carried to the HTL and then gathered by metal electrodes simultaneously. Finally, the FTO is connected to the metal electrodes to generate the photocurrent in the outer circuit.

3.5.4 ELECTRON TRANSPORT LAYER (ETL)

The electron transport layer (ETL) is a layer where its electrons exhibit high mobility and affinity. The primary purpose of the ETL in PSCs is to facilitate contact with the layer that harvests sunlight. This enhances the efficiency of photo-generated electrons to be extracted. As such, the ETL is a crucial element in PSCs. Different n-type semiconductor materials and structures have varying effects on the PCE.

When selecting the ETM for PSCs to operate more effectively, there are several aspects that take priority. First and foremost, n-type semiconductors with high carrier mobility offer advantages in providing higher efficiency for PSCs. Additionally, the ETM should allow penetration by sunlight, enabling PSCs to have a wide bandgap. Furthermore, the conditions for the preparation of PSCs should be gentle, and the ETM should be in a cold environment. Finally, the band structure should be compatible with PSCs to prevent errors and enhance performance. However, it's important to note that the ETM is not the main factor restricting efficiency.

Several aspects need investigation to strengthen the performance of the ETL, such as modifying the chemicals through doping and applying lamination to change the microstructure. These modifications enhance the electron transport within the PSCs. Laminated ETL improves the stability of PSCs and perovskite stability under UV illumination. For instance, the Mg-doping of a TiO_2 ETL improved the conduction band minimum alignment of the perovskite. Mg-doping of a SnO_2 ETL is ready to reduce the occurrence of cracking, increase the regularity and flatness of the ETL, and enhance the interfacial contact with the perovskite layer.

3.6 GRAPHENE

A 2D one-atom-thick sheet is known as graphene, a sp^2-bonded carbon material organized in a hexagonal pattern, serving as the building blocks of graphite. Graphene is unexpectedly strong, with the thinnest layer of material being 200 times stronger than steel. It has proven to possess excellent electrical, optical, and mechanical properties, along with good stability, contributing to the enhancement of the performance of PSCs. Graphene is also proficient at absorbing light and is a good conductor of heat and electricity. Much research has been conducted on graphene in the PV industry due to its unique characteristics, resulting in an increase in the efficiency of solar cells. The fusion of graphene into PSCs has been introduced, and extensive research has been carried out on this topic recently. This research covers various aspects, including holes and electrons transporting medium, electrodes, and so on, to ensure the devices become more stable. Figure 3.7 illustrates the structure of graphene.

Graphene is an extremely versatile material that can compound with other materials, such as gases and metals, to produce a different material with strong properties. Researchers worldwide are investigating its properties and applying them in many applications to benefit human life in this new era. The first application of graphene is its use as an essential material in batteries, such as solar cells. It is also employed in transistors, computer chips, antennas, water filters, supercapacitors, touchscreens for LCDs, and many more. Consequently, graphene has become a hotspot for researchers, especially Konstantin Novoselov and Andre Geim, who won the Nobel Prize in physics in 2010 as they were the first to isolate graphene in 2004 (Novoselov et al., 2004).

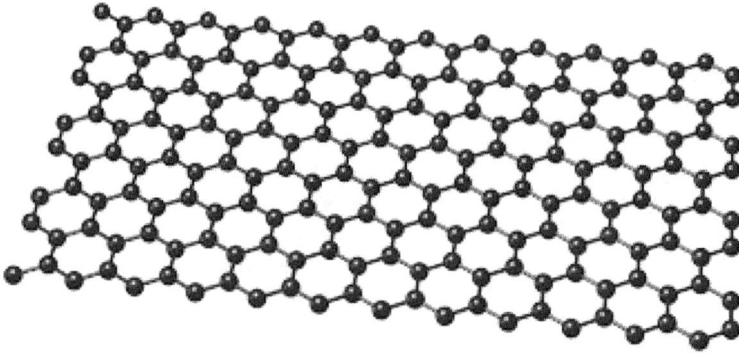

FIGURE 3.7 Structure of graphene.

3.6.1 Physical/Mechanical Properties of Graphene

Graphene is a material known as the thinnest in the world, measuring approximately 0.34 nm, and one of the toughest materials, being stronger than steel and diamonds in the same dimensions. Graphene can withstand more strength when stretched or pulled compared to other materials, as its tensile strength is even more than 1 TPa. Carbyne, a string of carbon, is the only material stronger than graphene. It also has a lightweight of only 0.77 mg/m². For comparison, 1 m² of paper is 1,000 times heavier than 1 sheet of graphene. Graphene also has the highest surface area since it exists in the 2D dimension. Graphene can stack to form the most stable structure of carbon called graphite. Moreover, graphene is water-resistant and cannot be penetrated by helium (Li et al., 2021).

Furthermore, graphene is excellent in conducting heat and has extraordinary potential for heat conduction based on its size. It has been found that the larger its size, the more heat it can conduct, as graphene has a stable bond pattern and a 2D structure. The thermal conductivity of graphene is expected to be around 3,000–5,000 W/mK at room temperature. Researchers believe that graphene's thermal conductivity is one of the highest compared to other materials. The high elasticity of graphene allows it to maintain its original size after being stretched. This is confirmed by atomic force microscopy for the spring constant with a thickness of 2–8 Nm, which is approximately 1–5 N/m. Its Young's modulus is approximately 0.5 TPa. This indicates that graphene can be stretched up to 20% of its starting size without damaging its structure.

3.6.2 Electronic Properties of Graphene

Graphene is a zero-overlapped semimetal, where its electrons and holes act as charge carriers. It is an excellent electrical conductor with high electrical conductivity. Carbon has six electrons, with two located in the s-orbit and the other four in the p-orbit. Only one of these electrons is freely available for electronic conduction while the atom is attached to the other three atoms. The high-mobility electrons are called pi electrons and contribute to enhancing the carbon bonds in graphene.

Holes and electrons in graphene do not have effective mass at the Dirac point, which is the six corners of the Brillouin zone. This is because the relationship between energy and movement is linear at low energy levels. Graphene has low electronic conductivity because there is no density of states at the Dirac point. Additionally, graphene exhibits high electric current density, a million times higher than copper. The resistivity of graphene is also lower than that of other materials at room temperature, and it can even transform into a superconductor, carrying electricity with 100% efficiency.

Graphene exhibits a high electron mobility, surpassing 15,000 cm²/Vs and having an upper limit of 200,000 cm²/Vs. However, this mobility is limited by the scattering of acoustic phonons around graphene. The mobility of graphene electrons is comparable to that of photons, as their mass is not significant enough to be measured. These charged carriers can travel sub-micrometers without scattering but are ultimately limited by the conditions of the substrate and the graphene itself. Despite being an excellent conductor, graphene is not suitable for making transistors because it lacks a bandgap.

Graphene is capable of harvesting 2.3% of white light, making it very suitable for manufacturing solar cells based on these properties. The unique characteristics of graphene are attributed to the high mobility of electrons. Researchers have demonstrated that the absorption of visible light is dependent on the Fine Structure Constant and not specific to the material. An interesting observation is that the amount of white light absorbed will be doubled if another layer of graphene is added to the first sheet. However, as the intensity of collected light reaches a threshold, the absorption of light will be reduced.

3.6.3   GRAPHENE-BASED ELECTRON TRANSPORT LAYER

As mentioned in Section 3.5, the ETL plays an essential role in PSCs. To enhance the PCE and stability for long-term performance, allowing efficient charge injection or collection at the electrodes, graphene-based or other 2D materials have been introduced into the structure of PSCs recently. Researchers have experimentally studied the effect of graphene-based ETLs in improving methylammonium lead iodide solar cells, leading to a significant enhancement in PSCs' efficiency.

The intersection between the ETL and perovskite is productively affected by the charge recombination process and material instability in PSCs. To slow down perovskite degradation and reduce nonradiative recombination, free charges are rapidly inserted from perovskite to ETL. Interface studies show that the conversion efficiency and stability are greatly enhanced when graphene flakes are inserted into the lithium-neutralized graphene oxide (GO-Li) and the mesoporous TiO_2 layer ($mTiO_2$) as an inter-layer at the perovskite and $mTiO_2$ interfaces. Hence, the effect of the substrate modified by mesoscopic graphene must be investigated in more detail to control the PSCs' performance in the PV industry.

Low-temperature graphene is often doped with tin oxide for planar PSCs. Doping graphene into the SnO_2 ETL, connected with the disabled charge fusion at the ETL and perovskite interfaces, has gradually enhanced efficiency. The graphene-SnO_2 device was found to display a higher efficiency of more than 18%. The device's J–V hysteresis is surprisingly attenuated, usually corresponding with the extraction of the electrons and enhanced mobility of electrons through the combination of the graphene conducted in two-dimensional electronics in the SnO_2 matrix. The graphene-SnO_2 improves stability, securing performance around 90% after exposure to air for 300 hours due to the graphene's hydrophobic nature.

There are major advantages to using graphene as a sandwich surface layer on the ETL to strengthen the intrinsic semiconducting properties of the ETL. It also transforms the surface and volume morphologies of the ETL. Through the utility of graphene with organic or inorganic molecules and chemical surface modification, the efficiency of graphene in the ETL of PSCs can be enhanced. As such, the surface chemistry and semiconducting properties can be measured so that it can adapt to different ETLs and PSCs. As graphene is stable in chemicals and possesses good physical properties, it enhances the stability of PSCs. Besides that, graphene can be integrated relatively easily into different conventional ETL materials.

3.6.4 DEFECTS OF GRAPHENE

Scientists have extensively investigated graphene since its discovery and found that lattice imperfections in graphene cannot be prevented during its growth. The chemical and electronic properties of graphene are affected by the defects in its structure. Therefore, it is essential to understand the defects in graphene. In this project, the aspect of defects needs to be introduced, including the types of defects in graphene. Additionally, solutions for addressing these defects in graphene must be detailed to improve the PCE performance.

Graphene, with its honeycomb network structure consisting of two carbon atoms per cell, has two categories for ribbon edges: zigzag and armchair. These ribbon edges can impact the electronic properties of graphene, similar to carbon nanotubes. However, a weakness arises when graphene is used as the ETL of PSCs due to the absence of a semiconducting gap in graphene. This lack of a bandgap makes it challenging for a device produced with graphene, as it lacks the capability to switch off. It also poses a disadvantage in the low static power consumption of complementary metal-oxide semiconductor (CMOS) technology.

Structural defects degrade the efficiency of PSCs with graphene-based ETLs, as graphene's outstanding mechanical and electronic properties only manifest in highly perfect conditions. However, imperfections may have utility in certain conditions, as they can lead to new applications and tailor the local properties of graphene. Structural defects in graphene can be induced through chemical treatment or irradiation.

Rings with non-hexagonal shapes may enable graphene to remain flat when symmetry rules are satisfied, offering deflection in the graphene sheet. Sp^2-hybridized carbon atoms can align into different polygons to form various architectures. Unlike other massive crystal-like materials such as silicon, graphene utilizes these properties. The atomic network is reorganized to allow a consistent network defect, eliminating under-coordinated atoms. These reorganized defects increase the reactivity of the framework and enable the absorption of other atoms onto the graphene.

3.6.5 Type of Defects in Graphene

Recently, scientists have been studying the structural deficiencies in carbon nano-tubes and graphite. Studies suggest that graphene should have deficiencies at the atomic level. However, accurately identifying all types of structural defects in graphene is challenging. High-resolution transmission electron microscopes (TEM) are employed to observe graphene at the atomic level, even down to single-layer graphene. Additionally, atomic force microscopes (AFM) and scanning electron microscopes (SEM) are experimental devices used to characterize nanomaterials and directly image configurations predicted theoretically.

There are two categories of defects in graphene: intrinsic defects and extrinsic defects. Intrinsic defects are characterized by non-sp^2 orbital hybridization in graphene. The existence of intrinsic defects is mainly due to non-hexagonal rings formed by surrounded hexagonal rings. The presence of non-carbon atoms disrupts the crystalline order in graphene. Studies on rebuilding the structure of carbon nanotubes under external energy disturbances indicate that the assumptions about defects may not always be stationary and can be random under certain conditions. They may move, governed by temperature and activation carriers.

Graphene has five categories of intrinsic defects: Stone–Wales defects, multiple-vacancy defects, single-vacancy defects, line defects, and carbon adatoms. Stone–Wales defects result from the rotation of a single pair of carbon atoms, causing adjacent carbon rings to become pentagons or heptagons. Rotation of carbon atom pairs creates Stone–Wales defects without carbon atoms being added or removed. The energy associated with these defects can reach about 5 eV. Stone–Wales defects can be induced by irradiation or rapid cooling in a high-temperature environment. Additionally, this defect may be formed by the impact of electrons. Figure 3.8 illustrates the atomic structure of Stone–Wales defects.

Graphene undergoes single-vacancy defects when one carbon atom is missing from the hexagonal rings. Additionally, graphene experiences Jahn–Teller distortion, causing two or three suspended bonds to connect at the missing atom to reduce its total energy. Geometry reasons may be attributed to the remaining pendant bonds. The energy produced by the vacancy defect is expected to reach 7.5 eV, demonstrating that higher energy than Stone–Wales defects is required to produce a vacancy defect with a pendant bond. Figure 3.9 illustrates the atomic structure of single vacancies in graphene.

The multiple-vacancy defect is also a structural defect in graphene, occurring when two or more carbon atoms are missing from the hexagonal rings (see Figure 3.10). Consequently, it can be inferred that the multiple-vacancy defect is based on the single-vacancy defect. For example, a double-vacancy defect is formed when two carbon atoms are missing from the ring. Double-vacancy defects result in the absence of pendant bonds for the rings formed by two pentagons and one octagon, instead of four hexagons. The energy required to form a double-vacancy defect is approximately 8 eV, and various numbered vacancy defects may form with different energy levels. This defect involves the rotation of carbon atoms, making the system more stable due to a reduction in total energy. Figure 3.10 illustrates the atomic structure of various multiple-vacancy defects.

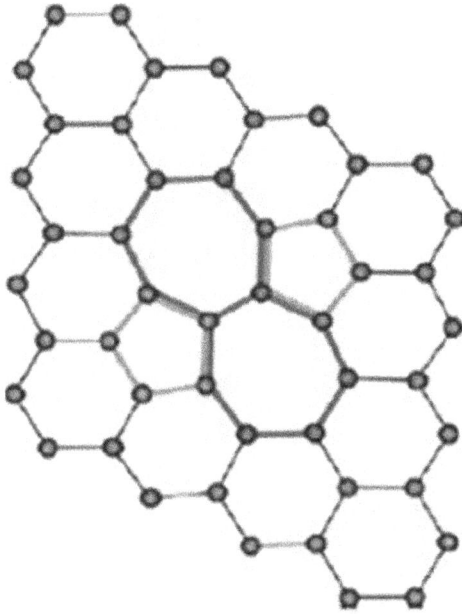

FIGURE 3.8 Atomic structure of Stone–Wales defects.

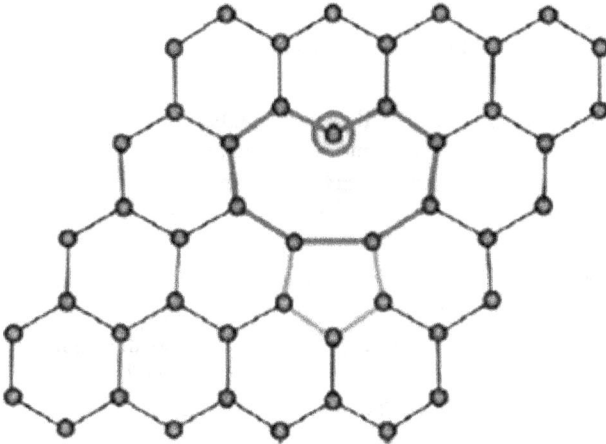

FIGURE 3.9 Atomic structure of single vacancy in graphene.

Another defect in graphene is the one-dimensional defect, also known as line defects, as shown in Figure 3.11. The line defects are slant borderlines dividing two regions of different networks, while the axis is perpendicular to the plane. Line defects can be considered as fixed-point defect lines that either have applied or not applied pendant bonds. One reason for the formation of these network regions is the mismatch of the lattice in graphene developed on the surface of Ni. These defects form due to octagons that divide the alternate lines of pentagon pairs.

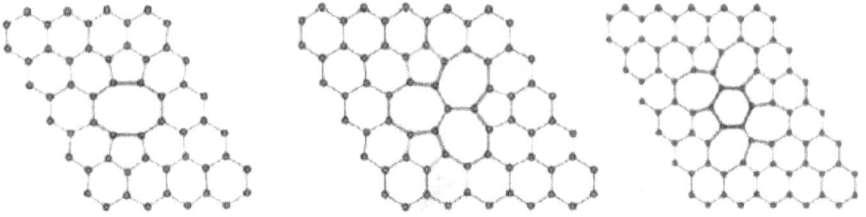

FIGURE 3.10   Atomic structure of various multiple-vacancy defect.

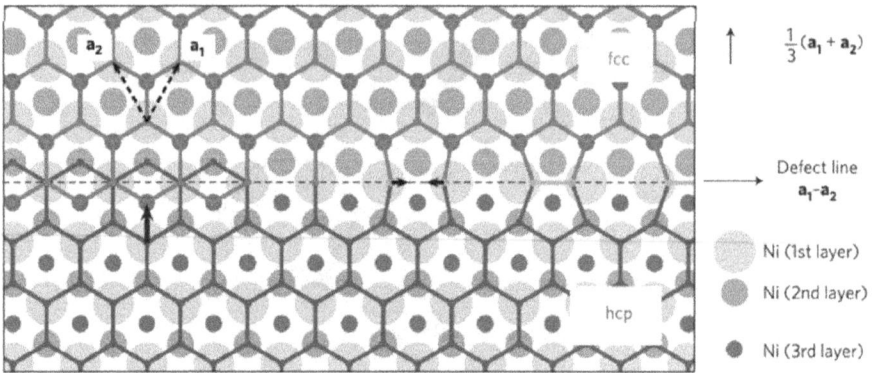

FIGURE 3.11   Structure of line defect in graphene.

   The final defect in graphene is the out-of-plane carbon adatoms. This defect is caused by missing carbon atoms that are not separated from the sheet of graphene when single or multiple-vacancy defects occur. These carbon atoms shift on the graphene's surface after breaking away from the initial hexagonal ring. A new bond is created when the carbon atoms shift to a new position in the plane. A new defect will form if the missing atoms come into contact with a new graphene sheet, leading to the destruction of the sheet's structure. This interaction may also result in the creation of a 3D structure in the graphene. These defects have a high migration rate and can disrupt the 2D crystal form of graphene. Figure 3.12 illustrates the atomic structure of out-of-plane carbon adatoms in graphene.

3.6.6   SOLUTION FOR GRAPHENE DEFECTS

Graphene defects are known for scattering phonons and charge carriers, reducing the length of the ballistic transmission path. The mobility of carriers and heat conductivity in graphene are negatively affected by the scattering of phonons and charge carriers. This negative impact of defects is particularly pronounced in graphene. For example, defects reduce the mobility of charge carriers produced by micromechanical cleavage. The defects are responsible for the properties of graphene films obtained by chemical methods, such as the chemical reduction of graphene oxide

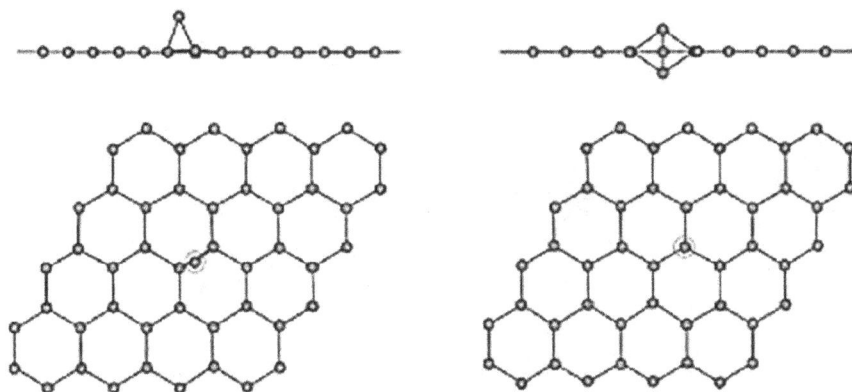

FIGURE 3.12 Defects and location of the carbon atoms (Tian et al., 2017).

(GO) and exfoliation. Therefore, solutions are needed to prevent these defects from producing high thermal and electrical conductivity nanostructures of carbon and, consequently, strengthen the mechanical strength.

First, thermal annealing is one of the solutions for defects by removing defects from the crystalline network. This process involves high temperatures in the presence of a hydrocarbon gas to restore the structure of graphite. Hydrocarbons break down to generate carbon atoms that can regenerate defective sites under certain conditions. López and his team recently introduced the processing of chemical vapor deposition (CVD) for chemically derived graphene films. They used ethylene carbon as a source to improve the film's conductivity by an order of magnitude ranging between 10 and 350 S/cm at room temperature. The enhancement of defective sites on the surface of graphitic material under conditions with hydrocarbon gas was observed and reported by Liu and his team. They investigated defect reactivity on the surface of highly oriented pyrolytic graphite (HOPG) at increasing temperatures using scanning tunneling microscopy (STM).

Self-healing is also a solution to remove defects in graphene. Tsetseris's team found that carbon adatom pairs and clusters of four or more self-interstitials remain hollow unless the system's temperature is elevated significantly. The clustering of three carbon adatoms causes the removal of hillock-like features, and the creation of mobile species causes the structure to self-heal. Liu's team demonstrated that most single-vacancy defects are likely to combine into larger vacancy holes, while the vacancies accumulate into a single hole in a layer through intra- and inter-layer migration in multi-layer graphene. Cooperation among defects simplifies the inter-layer migration of vacancies. Figure 3.13 illustrates the migration of vacancies between graphene layers, leading to the formation of vacancy holes in one layer and the self-healing of other layers.

Healing by absorption is also a method to reduce defects in graphene. Wang and his team introduced a method to control vacancy healing and N-doping in graphene. To heal the vacancy, CO and NO molecules need to be exposed sequentially. CO molecules are easily absorbed into the defective sites in graphene, while NO will

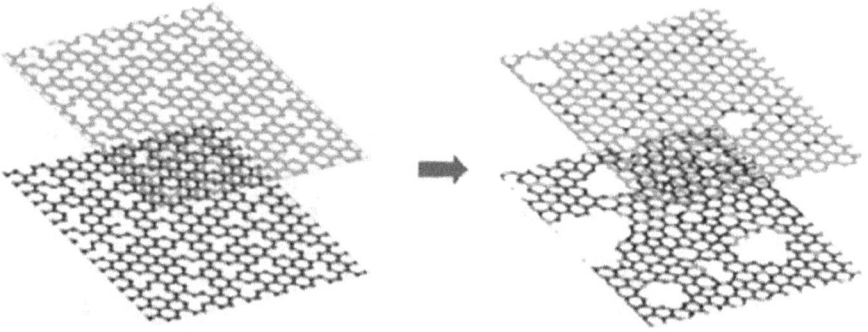

FIGURE 3.13 Vacancy migration between graphene layers leads to vacancies hole formation in one layer and the self-healing of other layers (Liu et al., 2014).

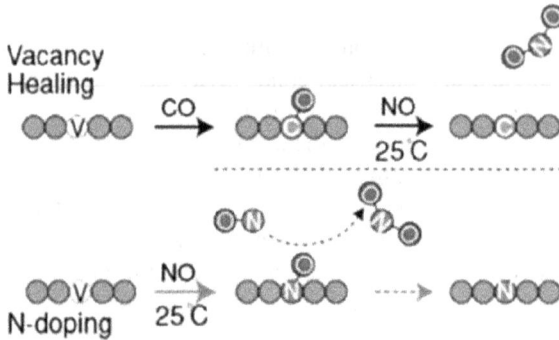

FIGURE 3.14 Vacancy healing and N-doping process of graphene with vacancies by CO and NO molecules (Wang and Pantelides, 2011).

chemically remove the additional oxygen atom to form NO_2 molecules. NO_2 molecules attach weakly to graphene, making them easy to extract from the material. This observation prompted the team to delve deeper into this topic, following the same steps until reaching an understanding of NO molecules. Therefore, the combination of CO and NO probably can facilitate synchronous healing and doping at a preferable ratio. This modulation is crucial for graphene in adjusting its bandgap. Figure 3.14 illustrates the vacancy healing and N-doping process of graphene with vacancies using CO and NO molecules.

Metal-assisted healing is also one of the solutions to remove defects in graphene. Karoui and his team investigated the healing of graphene developed on a metallic substrate using tight-binding Monte Carlo simulations. An isolated graphene sheet can become hardened, suggesting that its curing is thermally activated between 1,000 and 2,500 K. An excellent layer of graphene can be obtained when nickel is present as the substrate. The apparent structure indicates that the chemical bonds between nickel and carbon molecules keep breaking and reforming around the defective carbon zone. This process contributes to the primary communication necessary

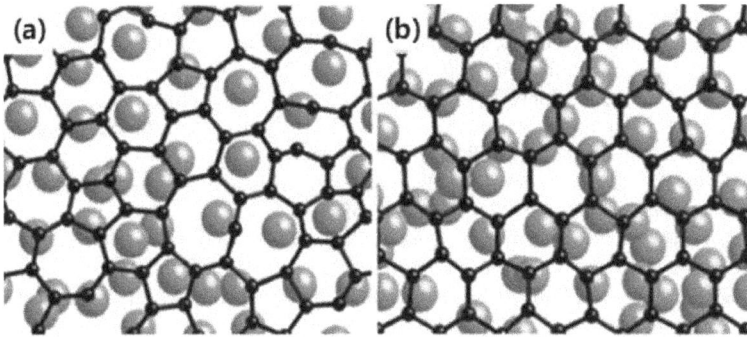

FIGURE 3.15   Defects at (a) 1000 K and (b) 2500 K (Karoui et al., 2010).

for healing. Significant defects are observed at a temperature of 1,000 K. Hexagonal and pentagonal rings are canceled out at high temperatures, leading to the healing of larger rings between 1,000 and 1,500 K. The graphene sheet is perfectly cured at 2,500 K. Figure 3.15 illustrates the defects at 1,000 and 2,500 K.

3.7   CONCLUSION

In conclusion, the defect density is unacceptable because it will influence the efficiency, performance, and stability of the PSCs. Additionally, heat dissipation is also a main factor for device thermal stabilization, with the help of a graphene layer. However, graphene has low conduction at accurate positions, and hence, it is suggested that graphene be doped with metallic ions. Furthermore, the interface of the graphene/perovskite layer must be set up with caution; otherwise, the recombination rate will be increased due to the huge network mismatch. The effect of the defect is also indicated as the transfer length (λ), in which the defect is expressed as the lengthening of the conductance channel. It is built upon the sheet resistance of graphene, and this amount contributes to a direct judgment for the PSC cell.

ACKNOWLEDGMENT

This research work was financially supported by the Fundamental Research Grant Scheme, FGRS (FRGS/1/2022/TK08/UTAR/02/39) from the Ministry of Higher Education and UTAR Research Fund, and UTARRF (IPSR/RMC/UTARRF/2022-C1/L12) under Universiti Tunku Abdul Rahman.

REFERENCES

Blume, E. (n.d.). Product review: PV technologies on the way to market. *Electric Perspectives,* 18:2.

Chapin, D.M., Fuller, C.S., and Pearson, G.L. (1954). A new silicon p-n junction photocell for converting solar radiation into electrical power. *Journal of Applied Physics* 25, 676–677.

Chen, C. (2023). Optimization strategies for electron transport layer aiming at high efficiency perovskite solar cells. *Journal of Physics: Conference Series* 2608, 012018.

Conibeer, G., Green, M., Corkish, R., Cho, Y., Cho, E.C., Jiang, C.W., Fangsuwannarak, T., Pink, E., Huang, Y., Puzzer, T., et al. (2006). Silicon nanostructures for third generation photovoltaic solar cells. *Thin Solid Film*s 511–512, 654–662.

France, R.M., Geisz, J.F., Song, T., Olavarria, W., Young, M., Kibbler, A., and Steiner, M.A. (2022). Triple-junction solar cells with 39.5% terrestrial and 34.2% space efficiency enabled by thick quantum well superlattices. *Joule* 6, 1121–1135.

Fritts, C.E. (1883). On a new form of selenium cell, and some electrical discoveries made by its use. *American Journal of Scienc*es 3, 465–472.

Karoui, S., Amara, H., Bichara, C., and Ducastelle, F. (2010). Nickel-assisted healing of defective graphene. *ACS Nano* 4, 6114–6120.

Kim, H.S., Seo, J.Y., and Park, N.G. (2016). Material and device stability in perovskite solar cells. *ChemSusChem* 9, 2528–2540.

Kojima, A., Teshima, K., Shirai, Y., and Miyasaka, T. (2009). Organometal halide perovskites as visible-light sensitizers for photovoltaic cells. *Journal of the American Chemical Society* 131, 6050–6051.

Li, X., Long, Y., Ma, L., Li, J., Yin, J., and Guo, W. (2021). Coating performance of hexagonal boron nitride and graphene layers. *2D Materials,* 3, 8.

Liu, L., Gao, J., Zhang, X., Yan, T., and Ding, F. (2014). Vacancy inter-layer migration in multi-layered graphene. *Nanoscale* 6, 5729–5734.

Novoselov, K.S., Geim, A.K., Morozov, S.V., Jiang, D., Zhang, Y., Dubonos, S.V., Grigorieva, I.V., and Firsov, A.A. (2004). Electric field in atomically thin carbon films. *Science* 306, 666–669.

Park, N.G., Grätzel, M., Miyasaka, T., Zhu, K., and Emery, K. (2016). Towards stable and commercially available perovskite solar cells. *Nature Energy* 1, 1–8.

Qian, J., Ernst, M., Wu, N., and Blakers, A. (2019). Impact of perovskite solar cell degradation on the lifetime energy yield and economic viability of perovskite/silicon tandem modules. *Sustainable Energy and Fuels* 3, 1439–1447.

Tian, W., Li, W., Yu, W., and Liu, X. (2017). A review on lattice defects in graphene: Types generation effects and regulation. *Micromachines* 8, 163.

Valsalakumar, S., Roy, A., Mallick, T.K., Hinshelwood, J., and Sundaram, S. (2023). An overview of current printing technologies for large-scale perovskite solar cell development. *Energies* 16, 1.

Wang, B., and Pantelides, S.T. (2011). Controllable healing of defects and nitrogen doping of graphene by CO and NO molecules. *Physical Review B: Condensed Matter and Materials Physics,* 24, 83.

4 Fuel Cells
Fundamental and Applications

Chiam-Wen Liew, Son Qian Liew, and Hieng Kiat Jun

4.1 INTRODUCTION TO FUEL CELL

Depletion of crude oil, natural gas, and coal catalyzes the development of the renewable energy source and planet-friendly technology (Viswanathan and Scibioh, 2007). Not only that, the concern about climate change exacerbates the development of alternative and sustainable energy sources. Among all electrochemical devices, fuel cell is a great choice to replace fossil fuels and oil because hydrogen can be an energy carrier source. Energy production using a clean technology becomes the core part of our daily life to solve the climate change issue. Fuel cell is an electrochemical device that converts the chemical energy of a fuel such as hydrogen to electrical energy through a redox reaction with minimized pollution (Wang et al., 2011; Kim et al., 2009). The concept of fuel cells was first introduced in 1839 by Sir William Grove, a Welsh judge and inventor, who discovered that electricity could be generated through the combination of hydrogen and oxygen. The first functioning fuel cell was built in 1843 by Sir Grove, who demonstrated that hydrogen and oxygen could be combined to produce electricity in a continuous and efficient manner. Fuel cell is a prominent electrochemical device as it exhibits a wide range of applications. Fuel cells have numerous advantages, such as being efficient, clean, safe, and reliable (Viswanathan and Scibioh, 2007). Among all the fuel cells, the proton exchange membrane fuel cell (also known as polymer electrolyte membrane fuel cell) (PEMFC) is a promising candidate. The PEMFC comprises of two porous gas diffusion electrodes (anode and cathode), proton conducting polymer electrolyte, anodic and cathodic catalyst layers (CLs), and current collectors. A great deal of effort has been made in the development of PEMFC as it has the potential to be an alternative energy source.

4.2 PRINCIPLES OF FUEL CELL OPERATION

4.2.1 MECHANISM OF PEMFCs

The working principle of a PEMFC is shown in Figure 4.8. The PEM utilized in PEMFCs is a membrane that separates the fuel (hydrogen) from the oxidant (air or oxygen), allowing the flow of the loaded molecules and ions (Giorgi et al., 2011). The catalyst used is mainly made of platinum and platinum-alloy nanoparticles, while the catalyst supports generally consist of high specific surface carbons such as volcanic

DOI: 10.1201/9781003314424-4

carbon, ordered porous carbon, hollow graphite particles, and more. Up to now, platinum with carbon-supported materials is the best candidate for PEMFC catalysts as Pt exhibits the most significant catalytic activity for the oxygen reduction reaction (ORR) to occur at the cathode and fuel oxidation at the anode side (Simya et al., 2018). The complex structure allows the electrochemical reactions to occur, where the transportation of reactant and product species, electrons, and protons from and to the active sites takes place. Moreover, the electrodes consist of anode and cathode, in which fuel oxidation and oxygen reduction occur, respectively. When hydrogen is used as fuel, protons and electrons are released due to oxidation at the anode. The protons generated are transported across the membrane, while the electrons are transported via the external circuit. At the cathode CL, the recombination of protons and electrons results in water generation (Weber et al., 2012). The electrochemical reactions involved in the operation of PEMFCs:

At anode: Fuel oxidation takes place.

- $H_2 \rightarrow 2H^+ + 2e^-$

At cathode: ORR takes place.

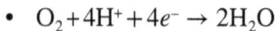

- $O_2 + 4H^+ + 4e^- \rightarrow 2H_2O$

Overall reaction

- $H_2 + 1/2O_2 \rightarrow H_2O$

Fuel cells operate in a continuous cycle, as long as hydrogen and oxygen are supplied to the electrodes. The basic working mechanism of a fuel cell can be summarized as follows:

1. First, hydrogen is supplied to the anode, where it is split into hydrogen ions and electrons.
2. The hydrogen ions pass through the proton exchange membrane to reach the cathode, while the electrons flow through an external circuit to generate an electrical current.
3. At the cathode, the hydrogen ions react with oxygen to form water, releasing heat in the process.
4. Finally, the electrical current generated by the flow of electrons can be used to power devices or be stored in a battery for later use.

4.2.2 Thermodynamics Principles of PEMFCs

Fuel cell is a type of electrochemical cell, where it converts chemical energy into electric energy. The chemical energy is usually supplied externally. During the energy conversion stage, water and heat are generated. The general layout of a fuel cell with its corresponding energy movement can be represented as in Figure 4.1.

In the case of a fuel cell system, it is an open system, where mass and energy flow through the system boundaries. As such, the work W is performed by the device on its surroundings. Thus, $-W$ over a period is to be expressed as

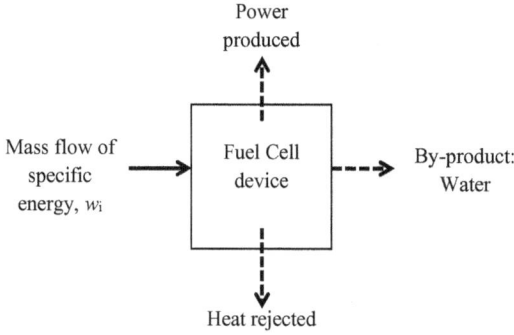

FIGURE 4.1   Schematic layout of a fuel cell with the respective energy movement.

$$-\Delta W = -\Delta W_{\text{electric}} + \int P \, dV \qquad (4.1)$$

where W_{electric}, P, and V are electric work, pressure, and volume, respectively. If the fuel cell system is at a fixed volume, $\int P \, dV$ is zero. This leaves us a pure energy output in the form of electric energy. This electric work is supplied by the fuel cell device to another device (or load) via an external circuit. Alternatively, electric work can be expressed in terms of the electric potential difference between the two connecting electrodes or terminals. In this relationship, the number of electrons involved can be quantified as

$$-\Delta W_{\text{electric}} = n_e \, N_A \, e \, \Delta\phi_{\text{ext}} \qquad (4.2a)$$

where n_e denotes the number of moles of electrons available, N_A is Avogadro's constant, and e represents the electron charge, thus $e\Delta\phi_{\text{ext}}$ becomes the energy difference for an electron. The product of $N_A \, e$ is Faraday's constant, F. Thus, Equation 4.2 can be simplified as

$$-\Delta W_{\text{electric}} = n_e \, F \, \Delta\phi_{\text{ext}} \qquad (4.2b)$$

Because fuel cell reaction involves a chemical reaction, Equation 4.2 can be rewritten to include the chemical potential difference, $\Delta\mu$ across the fuel cell device. This is the energy that can be absorbed or released during the reaction, and usually, it is made up by multiple components $\Delta\mu_i$:

$$-DW_{\text{electric}} = \sum_i n_{e,i} Dm_i \qquad (4.3)$$

In general, the chemical potential of component i (in this case, it's the electrolyte in the fuel cell device) can be expressed as

$$\mu_i = \mu_i^o + RT \ln a_i \qquad (4.4)$$

where R is the gas constant (8.3 J/K/mol), T is the temperature in Kelvin, a_i represents the activity coefficient in x_i of mole fraction, and μ_i^o is the standard potential of ith component.

From the definition of thermodynamics, the maximum amount of energy conversion by a fuel cell is represented by the Gibbs free energy.

$$G = U - T_{ref}S + P_{ref}V \qquad (4.5)$$

where U, T_{ref}, S, P_{ref}, and V are the internal energy of the fuel cell system, absolute temperature, entropy, pressure, and volume, respectively. These conditions are defined as they are in any static state situation. From the first law of thermodynamics, increment of internal energy is defined as the net energy added to the system from the surroundings. This relationship is expressed as

$$\Delta U = \int dQ + \int dW + \int dM \qquad (4.6)$$

where Q denotes the total net heat as supplied from the surroundings, W represents the net amount of work performed on the system by the environment, and M represents the net energy supplied to the fuel cell device. If we assume the fuel cell device to have a constant volume and the heat exchange with the surroundings is negligible, Equation 4.5 will become

$$-\Delta G = -\Delta W_{electric} \qquad (4.7)$$

which means the electricity produced is correlated with the loss of free energy from the fuel cell. In this instance, we can deduce that Equation 4.7 provides the maximum work that is extracted or converted between the system and the environment. By definition, the fuel cell device is in equilibrium thermodynamically when its free energy is zero.

With reference to Figure 4.1, the energy efficiency of the fuel cell device at a given instant in time is

$$\eta = (dQ/dt + dm/dt\ \Delta w)/(Q_i + dm/dt\ w_i) \qquad (4.8)$$

in which the energy flow rate is the result of the conversion of the mass flow rates by the specific energies ($\Delta w = w_i - w_o$). Ultimately, the device efficiency as defined by the first law of thermodynamics is simplified as

$$\eta = W/Q_i \qquad (4.9a)$$

where W is the work done given by the ΔG. In this case, heat input is calculated based upon the higher heating value (HHV) of the fuel.

$$\eta = \Delta G/HHV \qquad (4.9b)$$

Thus, the maximum efficiency will occur at standard conditions, with the highest possible cell voltage. For a PEMFC with hydrogen and oxygen as the fuel, the maximum calculated efficiency is about 83%. Although the efficiency seems high, it is impractical due to no current drawn (at open circuit conditions). Therefore, the limitation of the efficiency can be addressed by using the second law of thermodynamics. Thus, the efficiency of providing work is

$$\eta = W/W_{max} \qquad (4.10a)$$

$$\eta = \Delta G/\Delta G^* \qquad (4.10b)$$

$$\eta = n_e\, F\, E/n_e\, F\, E^o = E/E^o \qquad (4.10c)$$

where E is the system energy. The above equations are also known as voltage efficiency. It should be noted that the actual behavior of an energy conversion system device may not be accurately calculated due to various uncertainties with approximation. In any circumstances, direct application of the fundamental theories (as in this chapter) will serve as a guideline.

4.3 CATEGORY OF FUEL CELLS

Fuel cells are categorized by either their operating temperature (high temperature or low temperature) or the type of electrolyte used in the cell (for example molten carbonate salt is used in molten carbonate fuel cells – MCFCs). Fuel cells are categorized into five classes, i.e. alkaline fuel cells (AFCs), solid oxide fuel cells (SOFCs), phosphoric acid fuel cells (PAFCs), MCFCs, and PEMFCs, as listed in Figure 4.2.

4.3.1 ALKALINE FUEL CELLS (AFCS)

AFC is the firstly working fuel cell to generate electricity from hydrogen sources which was invented by Sir William Grove in the middle of the 19th century and developed into a workable system by Francis Bacon in the UK in the 1930s (Viswanathan and Scibioh, 2007). AFC is defined as the fuel cell that converts a controlled amount of hydrogen and oxygen into electricity using a circulating liquid alkaline electrolyte and two porous electrodes at a low operating temperature of 293–363 K. This cell usually consists of potassium hydroxide (30%–40% KOH) and two semi-porous electrodes with low platinum loadings of 0.3 mg/cm^2. The cell can

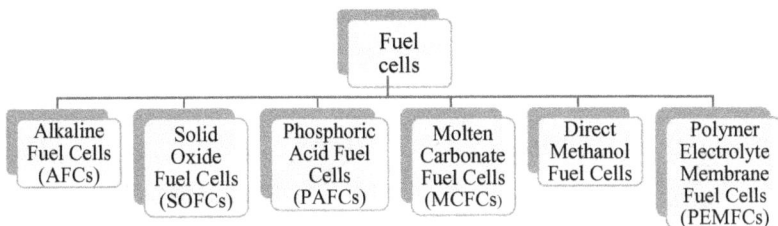

FIGURE 4.2 Types of fuel cells.

deliver power at ambient temperature. However, it just can produce full power at 343 K by applying electrical heating within 10 minutes of a cold start (Viswanathan and Scibioh, 2007). A schematic diagram of an AFC is shown in Figure 4.3.

An AFC cell is comprised of a potassium hydroxide (KOH) electrolyte and two porous electrodes. As shown in Figure 4.3, the oxygen from the air is supported to the positive electrode (cathode), whereas the hydrogen fuel is supplied to the negative electrode (anode). Anode and cathode are connected through an electric load. The voltage between these two electrodes of a single cell is between 0.9 and 0.5 V. There are two electrochemical redox reactions taking place at these respective electrodes. The reactions are illustrated as follows:

At cathode terminal: Oxygen is catalytically reduced to hydroxide anions.

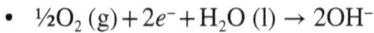

- $\frac{1}{2}O_2\,(g) + 2e^- + H_2O\,(l) \rightarrow 2OH^-$

At anode terminal: Hydrogen is catalytically reduced to water.

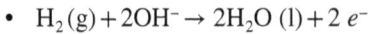

- $H_2\,(g) + 2OH^- \rightarrow 2H_2O\,(l) + 2\,e^-$

Overall reaction

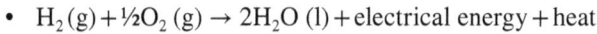

- $H_2\,(g) + \frac{1}{2}O_2\,(g) \rightarrow 2H_2O\,(l) + \text{electrical energy} + \text{heat}$

The AFC was initially designed to produce electrical power for mission-critical services in space applications. Therefore, it can be used as reliable power sources with low volume and light weight. The benefits of an AFC are also high efficiency, low cost, insensitive to ambient humidity, good reliability, and long cycle life. Furthermore,

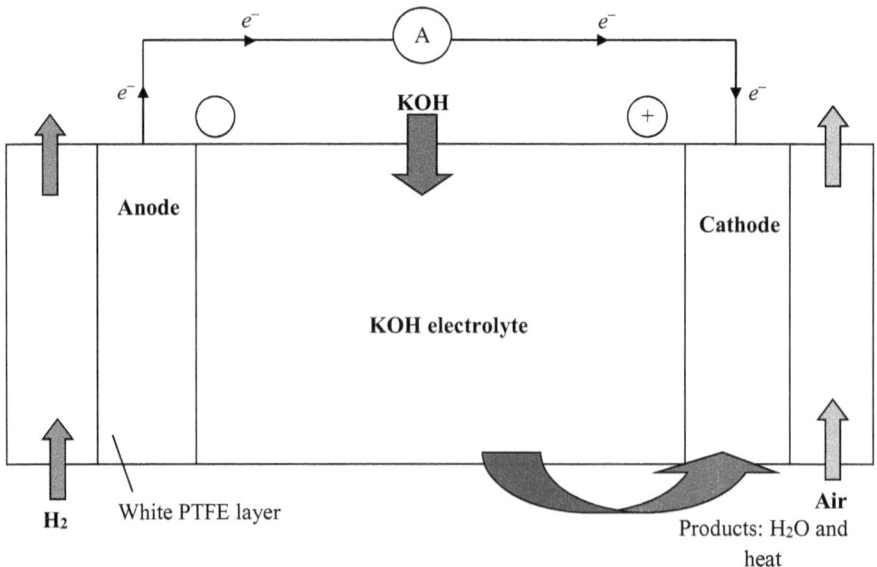

FIGURE 4.3   Schematic diagram of an AFC.

it does not rely on high-volume manufacture which can reduce the cost to an afford-able level. Fuel cells have the potential to enable a fast-refueling process for a variety of commercially available battery-powered electric vehicles. Nevertheless, there are several drawbacks associated with the use of fuel cells, including uneven catalyst distribution on porous electrodes, challenges in creating a precise pore system for reactant transportation, corrosive electrolytes, high susceptibility to catalyst polarization, and hazardous materials such as the use of asbestos in separators that can harm consumer health. Furthermore, air-breathing AFCs are intolerant to carbon dioxide (CO_2).

Modern and non-noble metal catalysts based on nickel, silver, and cobalt can be used to replace platinum to reduce the cost of the fuel cell system and improve the electrochemical performance of AFCs. However, nickel and silver catalysts could be degraded easily in the electrolyte which reduces the life span of the catalyst. AFC cannot be operated, especially in mobile applications if there is any CO_2 in the cathode feed gas streams. The potential problems of electrolyte contamination by atmospheric CO_2 can be completely overcome by using an atmospheric scrubber or by changing the electrolyte at the service intervals.

4.3.2 SOLID OXIDE FUEL CELLS (SOFCs)

SOFC is an electrochemical cell utilizing solid oxide or ceramics electrolyte, gaseous fuel, and oxidant to convert the chemical energy stored in fuel into electricity. SOFC is operated at high operating temperatures, ranging from 873 to 1,273 K which is much higher than AFC (Viswanathan and Scibioh, 2007). An SOFC is assembled with two electrodes (an anode and a cathode) and a ceramic electrolyte as a separator. The working principle of an SOFC is illustrated as follows (Figure 4.4):

Fuel (hydrogen, H_2 or methane, CH_4) is initially supplied to the anode. This fuel will react with the oxide ions from the electrolyte and, hence, release electrons (e^-) through an oxidation process. The released electrons are then transferred to the external circuit from the anode to the cathode. On the other side of the plate, oxidant (oxygen, O_2, or air) is fed to the cathode. As a result, the oxidants will accept the electrons from the external circuit and continue to supply the oxide ions (O^{2-}) for the electrolyte. Therefore, the oxygen ions conduction from the cathode and anode could maintain the overall electrical charge balance in the electrolyte. Hence, the

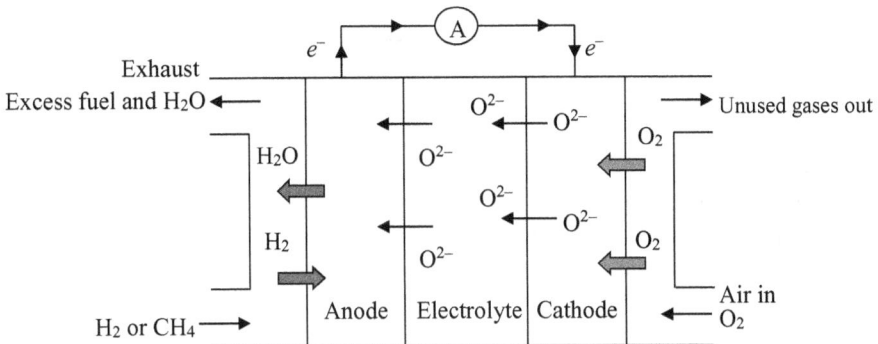

FIGURE 4.4   Schematic diagram of an SOFC based on oxygen-ion conductors.

electron flow and ion conduction will generate electricity from the combustion of the fuel and release the by-products of water and heat (Viswanathan and Scibioh, 2007; Stambouli and Traversa, 2002). The reaction at the anode and cathode of the cell is summarized as follows:

At anode terminal:

- $H_2 (g) + O^{2-} \rightarrow H_2O (l) + 2e^-$ or
- $CH_4 + 4O^{2-} \rightarrow 2H_2O (l) + CO_2 (g) + 2e^-$

At anode terminal:

- $CO (g) + O^{2-} \rightarrow CO_2 (g) + 2e^-$

At cathode terminal:

- $O_2 (g) + 4 e^- \rightarrow 2O^{2-}$

SOFCs have some merits over other types of fuel cells such as high system efficiency, high power density, and simple design (Viswanathan and Scibioh, 2007). This fuel cell does not require an external reformer as the reforming natural gas; other hydrocarbon fuels to extract necessary hydrogen can be accomplished within the fuel cell (Stambouli and Traversa, 2002). This fuel cell employs green technology because of its pollution-free feature. Since there is no combustion in the system, there will be no emission of acidic gas (carbon dioxide gas) and/or solid which results from the combustion (Mehmeti et al., 2016a). Apart from that, the need of platinum electrocatalysts can be eliminated (Viswanathan and Scibioh, 2007). High operating temperature of this fuel cell enhances reactant activity, enables fast electrode kinetics, and thereby reduces activation polarization (Stambouli and Traversa, 2002). Other features are reduced oil consumption, low operating and maintenance cost, high efficiency for electricity production, flexibility in choosing the fuel, and operating quietly because of solid state construction and long life cycle (40,000–80,000 hours) due to high tolerance to impurities such as sulfur and low CO_2 emission. In addition, high conversion efficiency in SOFCs compensates high cost of fuel. Another attractive point is this SOFC can be used as water electrolyzers at elevated temperatures without major modification. SOFC also offers high flexibility and easy setting up as a consequence of its modular properties (Stambouli and Traversa, 2002).

Although this fuel cell offers several advantages, it also has some demerits. This fuel cell relies on oxygen ion migration in the electrolyte. High purity of hydrogen is necessary in operating the fuel cell (Stambouli and Traversa, 2002). High electrolyte resistivity and electrode polarization might impede the transportation of oxide ions (Viswanathan and Scibioh, 2007). Formation of low conducting phase at the cathode–electrolyte interface is the main concern in developing this fuel cell. Apart from that, there is a limitation for scaling up SOFC power plants to megawatt size because of its brittle ceramic components. High capital cost-to-performance ratio inhibits the introduction of this fuel cell in the energy market. High operating temperature of SOFCs is the main contributor to affect the cell performance. The high operating temperature of this cell requires a significant startup time. The drop in operating temperature will decrease the performance of the cell. This is due to the increased internal resistance to the flow of oxygen ions. Moreover, an extra thermal shielding is required for this fuel cell to protect personnel and to retain heat because of high operating temperature (Viswanathan and Scibioh, 2007).

Some of the aspects of this fuel cell must be improved in order to enhance the performance of the cells, for instance, electrolyte preparation and homogeneity of materials, and compact fuel cell designs. The efficiency, durability, and lifespan of an SOFC are also the aspects to be investigated for research and development (Mehmeti et al., 2016b). Lowering the operation temperature can be a solution to build up a low-cost SOFC by employing a cheaper electrolyte such as ferritic steels without protective coatings of lanthanum chromate ($LaCrO_3$) and structural components such as stainless steel (Viswanathan and Scibioh, 2007). The reduction in operating temperature of the cell can increase the overall efficiency of the system and thus reduce the thermal stresses in active ceramic structures. Hence, this will expand the expected lifetime of the cell.

4.3.3 Phosphoric Acid Fuel Cells (PAFCs)

PAFCs were the first fuel cells for the commercialization as electric power sources in small- and medium-scale applications (Viswanathan and Scibioh, 2007). PAFC is a fuel cell employing concentrated liquid phosphoric acid (85–100 wt.% H_3PO_4) as the electrolyte and carbon paper coated with a finely dispersed platinum (Pt) catalyst as electrodes (Strickland et al., 2018). Hydrogen is extracted from fuels and then a positively charged proton migrates through the electrolyte from anode to cathode. Electrons are stripped off by the catalyst at anode and thus migrated to cathode through an external circuit. Direct current electric power of these fuel cells is eventually produced. The material of anode is usually polytetrafluoroethylene (PTFE, also known as Teflon) bonded with Pt/carbon black, whereas the material of anode is PTFE bonded with silicon carbide (SiC).

The schematic diagram of the working mechanism in a PAFC is illustrated in Figure 4.5.

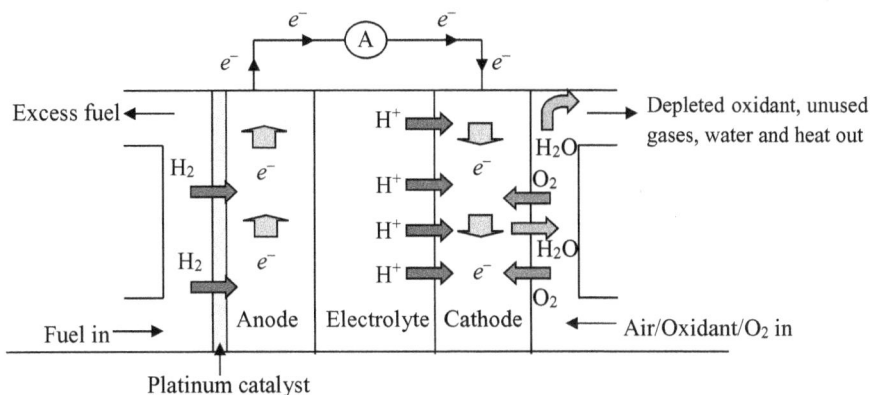

FIGURE 4.5 Schematic diagram of a PAFC.

The reaction occurred at both electrodes is listed as follows:
    At anode:

- $H_2 \leftrightarrow 2H^+ + 2e^-$

At cathode:

- $\frac{1}{2}O_2 + 2H^+ + 2e^- \leftrightarrow H_2O$
- $O_2 + 2H^+ + 2e^- \leftrightarrow H_2O_2$
- $H_2O_2 + 2H^+ + 2e^- \leftrightarrow 2H_2O$
- $H_2O_2 \leftrightarrow 2H_2O + \frac{1}{2}O_2$

These PAFCs exhibit several advantages and disadvantages. Higher overall efficiency which comes from heat and power (which is around 80%) is an attractive feature for developing PAFCs. Simple manufacturing process and cell construction are some of the advantages of PAFCs as these cells are mainly comprised of carbon black/carbon, PTFE, and SiC. Simple cell construction could lead to fabrication of tailor-made PAFCs. PAFCs also portray high total energy with excellent overall cell efficiency. Phosphoric acid is a preferred candidate due to its lower volatility property as the PAFCs operate at higher temperatures which is around 423–473 K above the boiling point of water for a long cycle life. It is due to other acid electrolytes requiring water for conductivity which cannot be operated at high temperatures. Phosphoric acid shows superior thermal, chemical, and electrochemical stabilities, excellent capillary properties, and low vapor pressure, even though it is not a good ionic conductor. Other inorganic acids such as hydrochloric, hydrofluoric, sulfuric, and perchloric acids have lower thermochemical stability and higher vapor pressure. Therefore, these inorganic acids are not suitable to be used as electrolytes in fuel cells. It is important to take note that sulfur must be removed if the hydrocarbon fuel is petrol because it will damage the platinum catalyst. PAFCs exhibit high tolerance toward a large variety of fuel gases, good thermal integration, and no CO_2 contamination as it is naturally rejected from the electrolytes.

    Platinum is used as catalyst at anode in PAFCs. However, rapid degradation of platinum is the main drawback of this type of FC. The degradation causes the loss in the surface area of the catalyst in the cell (Borup et al., 2007). The loss of surface area arises from dissolution of Pt at the electrode into the electrolyte at a cell voltage of more than 0.8 V and re-deposition of Pt to form larger particles via an electrochemical Ostwald-ripening mechanism (Strickland et al., 2018). In addition, these PAFCs deliver low specific power and power densities. Other limitations of PAFCs are slow rate of oxygen reduction, high material cost, precious metal required for the electrodes, and slow startup due to slightly higher operating temperature (Viswanathan and Scibioh, 2007).

    Due to their expensive cost, PAFCs are not considered a favorable choice. As a result, various research efforts have been dedicated to their development, with a focus on several aspects such as improving the resistance of the anode catalyst to poisoning, enhancing the efficiency of pressurization instruments, stabilizing the relationship between the cathode and catalyst, improving the interfacial stability between the electrode and electrolyte, developing a more durable catalyst material that can overcome Pt dissolution, increasing power density through the application of suitable electrolytes, and creating a cost-effective PAFC with high power and energy densities. These aspects have been studied extensively, as indicated by research (Viswanathan and Scibioh, 2007).

4.3.4 Molten Carbonate Fuel Cells (MCFCs)

MCFCs are fuel cells operating at high temperatures of 873 K and above. MCFCs are developed for natural gas– and coal-based power plants for electrical utility, industrial, and military applications. An MCFC consists of molten carbonate salt as an electrolyte and nickel as electrode material. The electrolyte can be lithium carbonate (Li_2CO_3) or potassium carbonate (K_2CO_3). On the other hand, the anode is made of nickel alloy, for instance, nickel alloyed with chromium (Cr) or aluminum (Al), whereas cathode is made of in situ lithiated nickel oxide (Mehmenti et al., 2016a). Figure 4.6 shows the basic configuration of an MCFC.

However, there are several shortcomings of MCFCs. The anode is prone to creep and may experience a decrease in surface area due to sintering. Additionally, the carbonate melt may not fully wet the anode. The cathode is susceptible to dissolution in the electrolyte and may experience a short circuit through the metallic Ni matrix. The matrix may crack due to thermal cycling and may lose its capillary retention of electrolytes. The electrolyte may corrode cell components and may also experience evaporation (EG and Technical, 2016; Kulkarni and Giddey, 2012).

4.3.5 Direct Methanol Fuel Cells (DMFCs)

Direct methanol fuel cells (DMFCs) are electrochemical devices that utilize methanol and methanol solutions as fuel, operating at near room temperature. Electrons and protons are discharged as the methanol and water are oxidized in the anode CL during operations, as shown in Figure 4.6. The electrons are delivered to the cathode by an external circuit, while the protons enter through the electrolyte membrane.

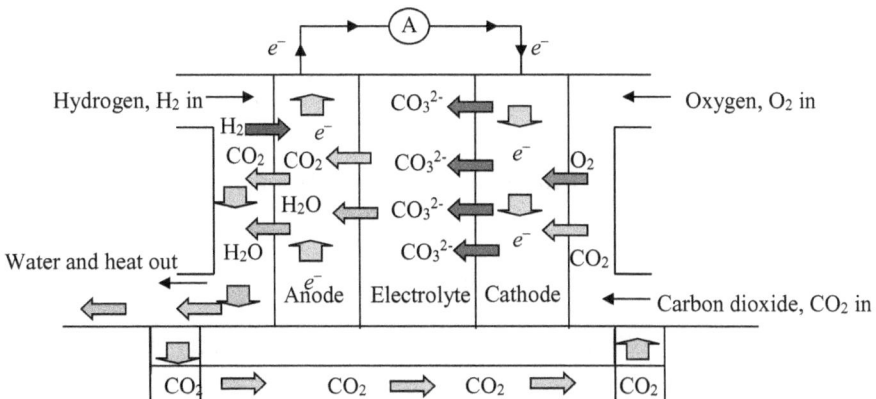

FIGURE 4.6 Schematic diagram of an MCFC.

When oxygen from the surrounding air reacts with the electrons and protons in the cathode CL, water is produced as a result (Li and Faghri, 2013) (Figure 4.7).

At anode CL: Methanol and water are oxidized to release electrons and protons.

- $CH_3OH + H_2O \rightarrow 6H^+ + 6e^-$

At cathode CL: Electrons and protons react with oxygen from atmospheric air, generating water.

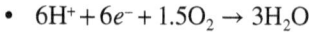

- $6H^+ + 6e^- + 1.5O_2 \rightarrow 3H_2O$

Overall reaction

- $CH_3OH + 1.5O_2 \rightarrow CO_2 + 2H_2O$

DMFCs are suitable for portable devices with high consumption of power. This is mainly owing to the easy handling of methanol and its high theoretical energy density (which is 6.1 kWh/kg per day). DMFCs may be classified as active, semi-passive, or passive based on ways of supplying fuel and oxidants. Active DMFCs use pumps, fans, sensors, heaters, and humidity components to supply methanol and oxidizing solutions. Fuel and oxidant supply rates can be precisely controlled, and the best performance can be achieved in active DMFCs. Passive DMFCs do not need any auxiliary methanol, water, and oxygen components. Capillary forces transfer the supplies, gravity, and gradients passively from the methanol tank and ambient air to the membrane electrode assembly (MEA). Passive DMFCs are more compact, reliable, and straightforward, and are therefore more suited to portable and low-power micro/miniature applications

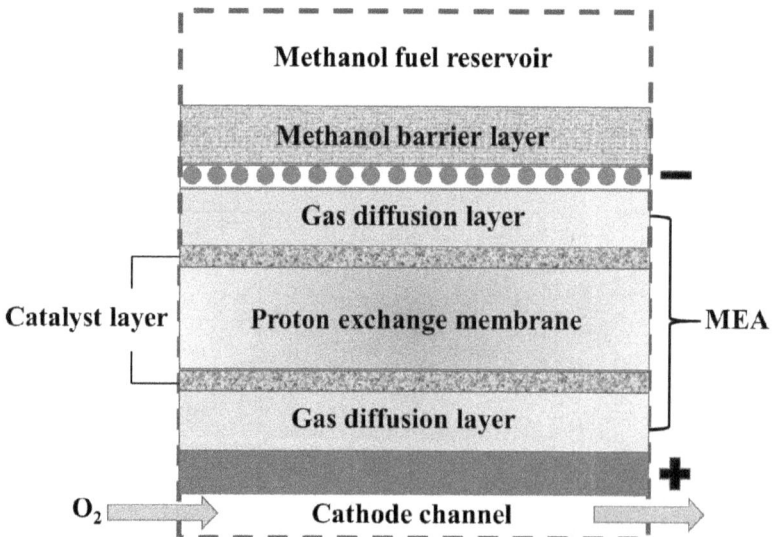

FIGURE 4.7 Semi-passive DMFC that is supplied with high concentration of methanol in the solutions. Adapted from Li and Faghri (2013).

(<10 W) (Verma, 2000; Liu et al., 2005; Faghri and Guo, 2008). In addition, semi-passive DMFCs have a passive and active anode or a passive cathode.

Nevertheless, passive DMFCs have specific barrier problems for commercialization purpose: methanol crossover (MCO), sluggish kinetic reaction of methanol, heat management, stability, durability, water management, cost, etc. (Munjewar et al., 2017).

4.3.6 POLYMER ELECTROLYTE MEMBRANE FUEL CELLS (PEMFCS)

PEMFCs, also known as proton exchange membrane fuel cells, have an anode, a cathode, and an electrolyte membrane. Hydrogen is oxidized at the anode, whereas oxygen is reduced at the cathode. An example of this process involves the transfer of protons through an electrolyte membrane from the anode to the cathode, while the electrons travel through external circuit loads. At the cathode, oxygen reacts with protons and electrons, resulting in the production of heat and water as byproducts (Spiegel, 2007). The whole PEMFC process is shown in Figure 4.8. In addition, depending on the working temperature, two distinct PEMFC types could be distinguished, where the first type is Low-Temperature Proton Exchange Membrane Fuel Cell (LT-PEMFCs), which works in the temperature range of 60°C–80°C, and the second type works in the range from 110°C to 180°C and is thus termed as the High-Temperature Proton Exchange Membrane Fuel Cell (HT-PEMFCs) (Sharaf and Orhan, 2014). Nafion, a Teflon-based polymer completely fluorinated and developed by DuPont in the 1960s

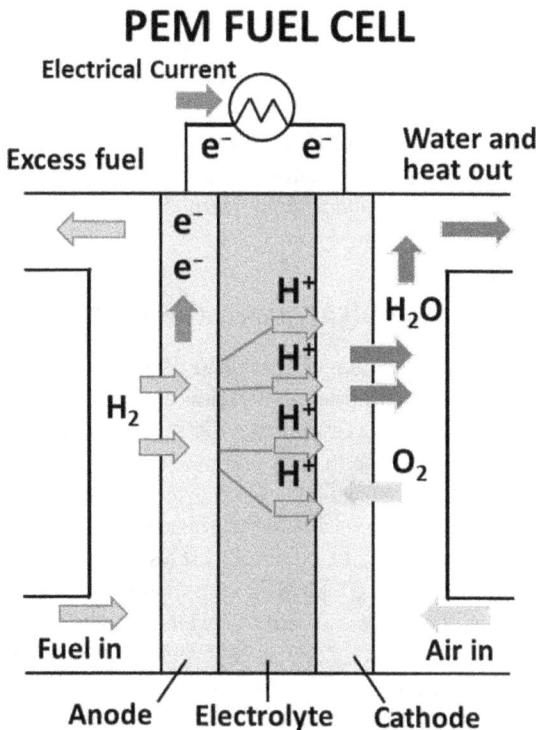

FIGURE 4.8   Schematic diagram of a PEMFC. Adapted from Mekhilef et al. (2012).

for space applications, is the usual electrolyte material employed in low-temperature PEM fuel cells. However, it is also possible to employ Nafion or polybenzimidazole (PBI) doped in phosphoric acid for HT-PEMFCs (Sharaf and Orhan, 2014). The traditional catalyst utilized in LT-PEMFCs at low temperatures is platinum, whereas Platinum–Ruthenium is utilized in catalysts for HT-PEMFCs (Sharaf and Orhan, 2014; Larminie et al., 2003). The electrical efficiency for LT-PEMs fuel cells is approximately 40%–60%, whereas for HT-PEMFCs, it is about 50%–60% (Sharaf and Orhan, 2014).

4.3.7 BIOFUEL CELLS

Biofuel cells produce electricity via electrochemical reactions which are catalyzed either by enzymes or by complete microbial cells (Davis and Higson, 2007; Bullen et al., 2006). Even though most organic substrates produce energy during combustion, oxidizers (for example, molecular oxygen, O_2 or hydrogen peroxide, H_2O_2) may be an alternative way of converting chemical energy into electrical energy by biocatalytic oxidization of organic substances, such as alcohol, organic acids, or sugars. Biocatalytic oxidation occurs explicitly at the anode, generating electrons that flow through the external electric circuit to another electrode (cathode) where the produced electrons serve as the source for the reduction reaction of oxidizer to occur. Organic fuel and oxidizer-aqueous solutions should circulate through the biofuel cell in which the bioelectrochemical processes occur to provide a constant power output. Biofuel cells utilize bioelectrocatalysts, such as redox enzymes or entire biological cells (e.g., microbial cells), which is different from fuel cells that use electrochemical inorganic catalysts (Luckarift et al., 2014; Luz et al., 2014; Hao and Scott, 2010; Das, 2018; Slate et al., 2018; Santoro et al., 2017; Du et al., 2007). A particular biofuel cell type is called "abiotic" biofuel cells that utilize inorganic catalysts to assist the oxidation process of bioorganic materials such as glucose. Unlike ordinary fuel cells, abiotic biofuel cells work under biologically compatible conditions (Katz and Bollella, 2021).

4.4 ADVANTAGES AND DISADVANTAGES OF PEMFCS

PEMFCs effectively convert chemical energy to usable electrical energy via electrochemical reactions and have garnered attention due to several advantages. These power sources have a greater maximum specific power per unit volume or weight than any other type of battery due to their compact design (Bagotskii, 2009). These cells operate at low temperatures, which is advantageous since it enables a fast startup procedure, but also disadvantageous due to the issue of sluggish reaction rates, which are handled via the employment of complex catalysts and electrodes (Sharaf and Orhan, 2014; Larminie et al., 2003). PEMFCs have been successfully used in a wide variety of applications. They are particularly well-suited for transportation applications owing to their advantages, such as high power density, rapid startup time, high efficiency, low operating temperature, less sensitivity to orientation, environmentally friendly, compact design, lightweight, solidity of electrolyte, flexibility in input fuel, and ease of handling. However, the PEMFC technology continues to face many obstacles such as high costs, poor durability, pure hydrogen, and oxygen required, water flooding, platinum catalyst poisoning, and hydrogen storage issues, all of which impede its commercialization.

4.5 APPLICATIONS OF FUEL CELLS

4.5.1 STATIONARY POWER GENERATION PLANTS

The PEMFC is the prevalent type of fuel cell used in stationary power applications, though there is increasing use of SOFCs, MCFCs, AFCs, and PAFCs (Carrette et al., 2001). Both low- and high-temperature fuel cells can be employed in stationary applications. While low-temperature fuel cells are typically more efficient during startup, high-temperature systems, like SOFCs and MCFCs, generate heat that can be used directly for other purposes and can also operate directly on fuels without external modification, resulting in higher efficiencies than low-temperature fuel cells (Carrette et al., 2001; Andújar and Segura, 2009). Figure 4.9 displays two main types of power stations: large stationary plants, which have a power output ranging from 300 kW to 20 MW, and small-to-medium stationary power plants, with power output ranging from a few Watts up to 10 kW for small stationary power plants, and from 10 up to 300 kW for medium stationary power plants (Das et al., 2021). Moreover, fuel cells are utilized for a variety of applications in stationary settings, serving as a primary source of power for the grid or remote-area power supply (RAPS) in locations where the grid is inaccessible. Furthermore, fuel cells are employed in hybrid power generation systems, in conjunction with photovoltaic systems, batteries, capacitors, or wind turbines, to provide primary or secondary power (i.e., distributed power or CHP generation), or to act as a backup power source (emergency backup power supply, EPS or UPS), depending on their characteristics (Spiegel, 2007; Sharaf and Orhan, 2014; Lucia, 2014).

By employing fuel cells, it is possible to transition from large-scale centralized power to decentralized distributed generation. Thanks to their static nature, minimal emissions, exceptional load tracking, and high efficiency, fuel cells can be used

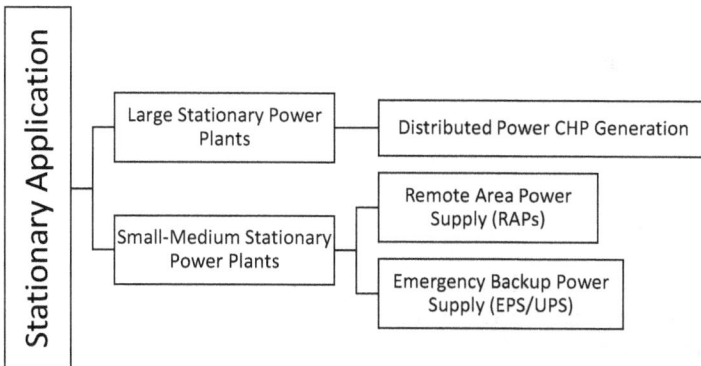

FIGURE 4.9 Applications of fuel cells in stationary power generation plants. Adapted from Alaswad et al. (2016).

to generate residential electricity or CHP power for households or larger residential blocks (Sharaf and Orhan, 2014). The combined heating and power systems boost energy efficiency by using both electrical and thermal energy. For instance, the produced heat can be used for space heating or different uses (Upreti et al., 2012). PEMFCs and PAFCs are used widely for CHP applications, especially for the household generation of CHP, but these fuel cells can only work at lower temperatures, typically less than 100°C and 200°C. Therefore, both types of fuel cells, SOFCs and MCFCs, require an external fuel reformer to use fuels other than pure hydrogen and have a low tolerance for CO, which is a byproduct of natural gas reforming (Zink et al., 2007). Therefore, high-temperature fuel cells, particularly SOFCs, are more appropriate for CHP generation on a larger residential block basis due to their characteristics (Sharaf and Orhan, 2014). At higher temperatures, SOFC systems have more considerable advantages than these technologies for the combined production of heat and electricity (Zink et al., 2007). Combined heat and power (CHP) fuel cell systems can be designed to function independently of the grid or with assistance from the grid. The grid-independent system is more expensive and complicated due to the need to handle dynamic load changes, but this can be resolved by addressing the fuel cell system size and its incorporation with battery banks or ultracapacitors. In contrast, the grid-assisted system exports electricity to the grid during periods of low demand and imports electricity during peak demand periods. In either scenario, an efficient CHP system necessitates a mechanism for storing heat (Sharaf and Orhan, 2014).

Medium-sized power stations are commonly utilized to provide combined electricity and heat to various structures, including cottages, administrative and office buildings, and hospitals. In contrast, small-power fuel cell power plants are constructed to supply electricity to individual residential or administrative areas (Bagotskii, 2009). PEMFCs are the most commonly used fuel cells for low-energy plants, such as the 1 kW power plant established by Ebara Ballard to produce CHP. This unit is designed to function for 10 years and comply with government standards. Similarly, the Japanese company Fuji Electric produced a similar device for 10 years of operation, with a cost of 12,000–16,000 dollars (Bagotskii, 2009). Fuel cells are now becoming a promising alternative to electric power system (EPS) batteries, particularly in the telecommunications market, with PEMFCs and DMFCs being the primary types of fuel cells selected due to their high-energy, high power densities, high modularity, extended operating times, compact sizes, and high reliability (Sharaf and Orhan, 2014). While an EPS requires a high degree of reliability, it does not necessarily require high operational durability. On the other hand, EPS has been proven to have successful applications in different sectors, including data centers, banks, government institutions, hospital surgical wards, traffic control computer systems, financial institutions, and emergency lighting systems for public spaces (Sharaf and Orhan, 2014; Bagotskii, 2009; Cacciola et al., 2001).

Fuel cells are utilized in RAPS where power generation can be challenging in grid-isolated locations such as islands, deserts, forests, remote technical installations, and research centers (Sharaf and Orhan, 2014). For instance, a fascinating initiative on Bechervaise Island in Antarctica, which includes hydrogen production through electrolysis using wind energy and low-power electrical and heat-producing fuel cell units, is funded by a $600,000 grant from the Australian Government (Bagotskii, 2009).

4.5.2 AUTOMOTIVE INDUSTRY

Nowadays, the automotive industry is one of the primary areas of clean energy technology development. This is because the transport sector accounts for 17% of the world's greenhouse gas emissions annually (Sharaf and Orhan, 2014). Therefore, significant optimizations are required from this industry to achieve the Kyoto Protocol's objectives (Cacciola et al., 2001). Investing in technology is crucial for addressing the concerns, particularly in reducing harmful emissions and improving energy conversion efficiencies. Fuel cells offer a prime example, as they can enable the automotive industry to achieve nearly zero hazardous emissions while maintaining the efficiency of the vehicle's propulsion system. In addition to that, fuel cells possess efficiencies nearly twice the efficiencies of traditional internal combustion engines. Due to its benefits of smooth operation, fuel flexibility, modularity, and low maintenance needs, the fuel cell is a promising candidate to eventually replace traditional combustion engines in the future. This is primarily because of the absence of moving parts. The fuel cell business currently faces several commercialization issues in the transportation/automotive sector. One crucial difficulty is the cost of the fuel cell, the unit's durability, and its performance. Other than that, hydrogen storage and hydrogen fuel infrastructure must all be considered (Alaswad et al., 2016).

PEMFCs possess a range of characteristics that make them ideal for transportation and automotive applications, including high-power density, high conversion efficiency, compactness, lightweight design, and low operating temperatures. Nevertheless, the use of fuel cells in automobiles is associated with several challenges, including issues with volumetric and gravimetric power density and reliability, as well as costs. Indeed, the high cost of the technology remains the most significant barrier to commercialization. Another promising fuel cell technology for transportation applications is DMFC, but it is not as advanced as PEMFC and has lower power densities due to methanol absorption through the membrane and catalyst toxicity. Apart from that, SOFC is another potential development technology in the field of transport applications. Different from PEMFCs, SOFCs allow the utilization of hydrocarbon fuels, even if they are complex to be used.

On top of that, they have no significant issue with the poisoning of catalysts. Numerous companies have developed SOFCs as auxiliary power units (APUs) for petrol automobiles due to these advantageous properties (Carrette et al., 2001). The application of fuel cells in the automotive industry is summarized in Figure 4.10 (Sharaf and Orhan, 2014).

Fuel cells have significant applications in light traction vehicles (LTVs), such as scooters, personal wheelchairs, motorcycles, airport tugs, golf carts, electric-assisted bicycles, and material handling vehicles (e.g., carts, pallet trucks, and forklifts). One of the most important potential applications of fuel cells in LTVs could be in motorcycles, as they are a significant contributor to pollution in urban areas despite their small size. The successful integration of fuel cells in the transportation industry can be seen in forklifts, with around 1,300 fuel cell-powered forklifts currently in use in the US market (Sharaf and Orhan, 2014). PEMFCs are the most often used fuel cells in research and development for light fuel cell electric vehicles (L-FCEVs) due to their benefits. FCEVs operate on a straightforward principle: The electricity is

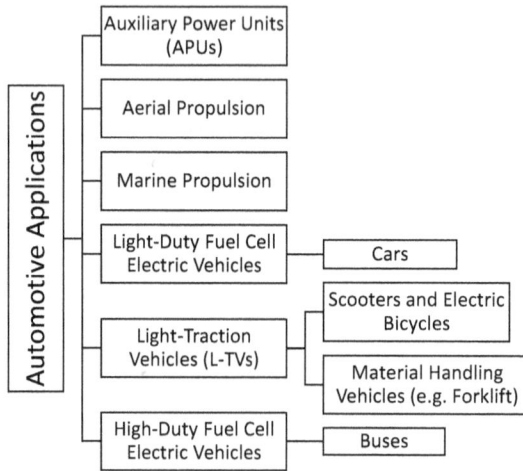

FIGURE 4.10 Automotive applications for fuel cells. Adapted from Sharaf and Orhan (2014).

created from hydrogen using low-temperature fuel cells (most commonly PEM), which are then utilized to power the vehicle or stored in batteries or ultracapacitors (Pollet et al., 2012). Since fuel cells create energy through chemical reactions, they do not emit toxic emissions (only generate water as a byproduct) and generate lesser heat compared to that of internal combustion engines. Despite the benefits of PEMFCs, many automobile manufacturers prefer methanol as a fuel source since it can be used to generate hydrogen on-board via steam reforming. However, the downside of this approach is longer response times and requires considerable gas cleaning before fueling the fuel cell (Alaswad et al., 2016). Due to these problems and the size of the reformer and gas purification unit, DMFCs are more ideal for mobile systems (Carrette et al., 2001). Many car manufacturers are putting much effort into developing L-FCEVs' commercialization (e.g., General Motors, Toyota, Mazda, Daimler AG, Volvo, Volkswagen, Honda, Hyundai, Nissan) (Sharaf and Orhan, 2014). For instance, the car manufacturer Honda introduced the FCX Clarity model at the automotive saloon in Los Angeles in 2007. The world's first fuel cell vehicle platform produced in series has been available in the market since the summer of 2008 (Sharaf and Orhan, 2014). At the 2010 Fuel Cell Expo in Tokyo, Japanese car manufacturers announced a plan to introduce two million FCEVs in Japan by 2025, which would make them completely competitive (Pollet et al., 2012).

The category of heavy-duty electric fuel cell vehicles (H-FCEVs) includes buses, heavy-duty trucks, locomotives, vans, utility trucks, service fleets, and other vehicles that utilize fuel cells as part of their high-capacity electric propulsion systems. In recent times, several fuel cell buses have been operating in different parts of the world (Andújar and Segura, 2009). In 2012, more than 30 buses that utilized fuel cells were deployed in Western Europe and 25 in the United States. The most common applications of fuel cell stacks in the industry are PEMFCs and PAFCs, which exhibit remarkable renewable braking energy recovery and excellent dynamic responsiveness. At present, fuel cell electric buses (FCEBs) are not economically competitive with conventional buses and other alternative technologies due to the advanced nature of fuel cell technology and the lack of mass production and manufacturing capabilities (Sharaf and Orhan, 2014).

Meanwhile, the use of fuel cells in marine transportation has been on the rise in recent years. While fuel cells offer several advantages for ships and ferries, including high efficiency, low emissions, and stationary operation, they face challenges related to reliability, durability, shock resistance, and the corrosive effects of saltwater air (Sharaf and Orhan, 2014). As a result, fuel cell applications in marine transportation have not been explored as extensively as they have in other sectors like automotive or aviation propulsion. On top of that, high cost, harsh marine environment with dynamic stresses and severe corrosion, life cycle, and lack of infrastructure are barriers that prevent the commercialization of fuel cells in this field. In Germany, the first commercial passenger ship with a hybrid PEMFC/battery system was demonstrated in 2003, followed by the world's first commercial passenger ship with the same system in 2008. Additionally, submarines that use oxygen/hydrogen fuel cells for propulsion and auxiliary power have been successfully deployed. The PEMFC technology is the preferred choice for providing air-independent propulsion systems (AIP) for a new generation of conventional submarines in several countries, including Germany, Greece, Italy, Portugal, and Spain (Leo et al., 2010).

The space industry was one of the first areas in which fuel cells were implemented. NASA deployed AFCs and PEMFC technology in the 1960s for the manned space programs. Fuel cells are considered promising for space applications due to their many advantages over other power generation technologies. Of particular interest is the fact that water is a byproduct of the electrochemical reactions that take place within the fuel cell, making it particularly useful for space missions where air, water, and food are crucial. However, due to the high energy density, power density, durability, and reliability requirements, it is currently not practical to commercialize the use of fuel cells in manned military and commercial air vehicles (Sharaf and Orhan, 2014). The summarized application of fuel cells in the transportation industry is shown in Table 4.1 (Alaswad et al., 2016).

4.6 CONCLUSIONS

In conclusion, fuel cell technology stands at the forefront of sustainable energy solutions, offering a versatile and efficient means of electricity generation. With a primary focus on hydrogen fuel cells, which seamlessly convert the chemical potential of hydrogen and oxygen into clean electric power and water, the applications span across diverse sectors including transportation, stationary power generation, and energy storage. Fuel cells offer a potentially non-polluting and renewable way to generate electricity. PEMFCs offer several advantages over the batteries and supercapacitors. Fabrication of a fuel cell with high power density, long durability, and cost-effective properties is essential for the economic application of sustainable energy technologies. The research and development of PEMFCs is still essential to harness the capabilities of hydrogen for energy production to the maximum threshold and operational advantages prior to commercialization. Innumerable challenges and technological breakthroughs are still the obstacles in the R&D of PEMFCs.

TABLE 4.1

The Applications of Fuel Cells in the Transportation Industry

Application		PEMFC	DMFC	SOFC	MCFC	PAFC	AFC	MFC
Stationary	CHP	#			#		#	
	RAPS	#			#	#		
	EPS	#		#	#	#		
Portable	Power generators	#	#					
	Consumer electronics	#	#					
	Battery charges	#	#					
	Army	#	#					
Transportation	APUs	#	#		#	#		
	LTVs	#	#					
	L-FCEVs	#						
	H-FCEVs	#					#	
	Aerial propulsion	#			#			#
	Marine propulsion	#			#	#		

ACKNOWLEDGMENT

This work was partly supported by the Fundamental Research Grant Scheme (FRGS/1/2019/STG07/TARUC/02/1) from the Ministry of Higher Education of Malaysia.

REFERENCES

Alaswad A, Palumbo A, Dassisti M, Olabi AG (2016) Fuel cell technologies, applications, and state of the art. In: *A Reference Guide in Reference Module in Materials Science and Materials Engineering*. Elsevier, Amsterdam.

Andújar JM, Segura F (2009) Fuel cells: History and updating. A walk along two centuries. *Renewable and Sustainable Energy Reviews* 13: 2309–2322.

Bagotskii VS (2009) *Fuel Cells: Problems and Solutions*. John Wiley & Sons, New York.

Borup R, Meyers J, Pivovar B, Kim YS, Mukundan R, Garland N et al. (2007) Scientific aspects of polymer electrolyte fuel cell durability and degradation. *Chemical Reviews* 107: 3904–3951.

Bullen RA., Arnot TC, Lakeman JB, Walsh FC (2006) Biofuel cells and their development. *Biosensors and Bioelectronics* 21: 2015–2045.

Cacciola G, Antonucci V, Freni S (2001) Technology up date and new strategies on fuel cells. *Journal of Power Sources* 100: 67–79.

Carrette L, Friedrich KA, Stimming U (2001) Fuel cells: fundamentals and applications. *Fuel Cells* 1: 5–39.

Das D (2018) Microbial fuel cell: A bioelectrochemical system that converts waste to watts. In: Microbial Fuel Cell: A Bioelectrochemical System that Converts Waste to Watts. Springer International Publishing. pp. 1–506. https://doi.org/10.1007/978-3-319-66793-5

Das HS, Chowdhury MFF, Li S, Tan CW (2021) Fuel cell and hydrogen power plants. In: Kabalci E (ed) *Hybrid Renewable Energy Systems and Microgrids*. Acedemic Press Inc, San Diego, pp. 313–349

Davis F, Higson SPJ. (2007) Biofuel cells-Recent advances and applications. *Biosensors and Bioelectronics* 22: 1224–1235.

Du Z, Li H, Gu T (2007) A state of the art review on microbial fuel cells: A promising technology for wastewater treatment and bioenergy. *Biotechnology Advances* 25: 464–482.

EG and Technical (2016) *Fuel Cell Handbook*, 7th edn. Department of Energy Office of Fossil Energy National Energy Technology Laboratory, United States.

Faghri A, Guo Z (2008) An innovative passive DMFC technology. *Applied Thermal Engineering* 28: 1614–1622.

Giorgi L, Salernitano E, Gagliardi S, Dikonimos T, Giorgi R, Lisi N, De Riccardis F, Martina V. (2011) Electrocatalysts for methanol oxidation based on platinum/carbon nanofibers nanocomposite. *Journal of Nanoscience and Nanotechnology* 11: 8812–7.

Hao YE, Scott K (2010) Enzymatic biofuel cells-fabrication of enzyme electrodes. *Energies* 3: 23–42.

Katz E, Bollella P (2021) Fuel cells and biofuel cells: From past to perspectives. *Israel Journal of Chemistry* 61: 68–84.

Kim DS, Cho HI, Kim DH, Lee BS, Yoon SW, Kim YS, Moon GY, Byun H, Rhim JW (2009) Surface fluorinated poly(vinyl alcohol)/poly(styrene sulfonic acid-co-maleic acid) membrane for polymer electrolyte membrane fuel cells. *Journal of Membrane Science* 342: 138–144.

Kulkarni A, Giddey S (2012) Materials issues and recent developments in molten carbonate fuel cells. *Journal of Solid State Electrochemistry* 16: 3123–3146.

Larminie J, Dicks JLA, Dicks A, Dicks AL, Knovel (2003) *Fuel Cell Systems Explained*. John Wiley & Sons, New York.

Leo T, Durango JA, Navarro E (2010) Exergy analysis of PEM fuel cells for marine applications. *Energy* 35: 1164–1171.

Li X, Faghri A (2013) Review and advances of direct methanol fuel cells (DMFCs) part I: Design, fabrication, and testing with high concentration methanol solutions. *Journal of Power Sources* 226: 223–240.

Liu J, Zhao T, Chen R, Wong C (2005) The effect of methanol concentration on the performance of a passive DMFC. *Electrochemistry Communications* 7: 288–294.

Lucia U (2014) Overview on fuel cells. *Renewable and Sustainable Energy Reviews* 30: 164–169.

Luckarift H, Atanassov P, Johnson GR (2014) *Enzymatic Fuel Cells: From Fundamentals to Applications*. John Wiley & Sons, New York.

Luz RAS, Pereira AR, de Souza JCP, Sales FCPF, Crespilho FN (2014) Enzyme biofuel cells: thermodynamics, kinetics and challenges in applicability. *ChemElectroChem* 1: 1751–1777.

Mehmeti A, McPhail SJ, Pumiglia D, Carlini M (2016a) Life cycle sustainability of solid oxide fuel cells: From methodological aspects to system implications. *Journal of Power Sources* 325: 772–785.

Mehmeti A, Santoni F, Pietra MD, McPhail SJ (2016b) Life cycle assessment of molten carbonate fuel cells: State of the art and strategies for the future. *Journal Power Sources* 308: 97–108.

Mekhilef S, Rahman S, Safari A (2012) Comparative study of different fuel cell technologies. *Renewable & Sustainable Energy Reviews* 16: 981–989.

Munjewar SS, Thombre SB, Mallick RK (2017) Approaches to overcome the barrier issues of passive direct methanol fuel cell: Review. *Renewable and Sustainable Energy Reviews* 67: 1087–1104.

Pollet BG, Staffell I, Shang JL (2012) Current status of hybrid, battery and fuel cell electric vehicles: From electrochemistry to market prospects. *Electrochimica Acta* 84: 235–249.

Santoro C, Arbizzani C, Erable B, Ieropoulos I (2017) Microbial fuel cells: From fundamentals to applications. A review. *Journal of Power Sources* 356: 225–244.

Sharaf O, Orhan M (2014) An overview of fuel cell technology: Fundamentals and applications. *Renewable and Sustainable Energy Reviews* 32: 810–853.

Simya OK, Nair P, Ashok A (2018) Engineered nanomaterials for energy applications: In: Chaudhery MH (ed) *Handbook of Nanomaterials for Industrial Applications*. Elsevier, Amsterdam, pp. 751–767.

Slate A, Whitehead K, Brownson D, Banks C (2018) Microbial fuel cells: An overview of current technology. *Renewable and Sustainable Energy Reviews* 101: 60–81.

Spiegel C (2007) *Designing and Building Fuel Cells*. Citeseer, New York.

Stambouli AB, Traversa E (2002) Solid oxide fuel cells (SOFCs): A review of an environmentally clean and efficient source of energy. *Renewable and Sustainable Energy Reviews* 6: 433–455.

Strickland K, Pavlicek R, Miner Elise, Jia Q, Zoller I, Ghoshal S, Liang W, Mukerjee S (2018) Anion resistant oxygen reduction electrocatalyst in phosphoric acid fuel cell. *ACS Catalysis* 8: 3833–3843.

Upreti G, Greene D, Duleep KG, Sawhney R (2012) Fuel cells for non–automotive uses: Status and prospects. *International Journal of Hydrogen Energy* 37: 6339–6348.

Verma L (2000) Studies on methanol fuel cell. *Journal of Power Sources* 86: 464–468.

Viswanathan B, Scibioh MA (2007) *Fuel Cells: Principles and Applications*. CRC Press, Taylor & Francis Group, UK.

Wang Y, Chen KS, Mishler J, Cho SC, Adroher XC (2011) A review of polymer electrolyte membrane fuel cells: Technology, applications, and needs on fundamental research. *Applied Energy* 88: 981–1007.

Weber AZ, Balasubramanian S, Das PK (2012) Proton exchange membrane fuel cells. In: Sundmacher K (ed) *Advances in Chemical Engineering*. Academic Press, San Diego, pp. 65–144.

Zink F, Lu Y, Schaefer L (2007) A solid oxide fuel cell system for buildings. *Energy Conversion and Management* 48: 809–818.

5 Solid-State Polymer Electrolytes for Fuel Cells Application

Chiam-Wen Liew, Son Qian Liew, and Hieng Kiat Jun

5.1 INTRODUCTION

Several electrochemical devices are emerged as energy storage devices and as power sources in recent years such as fuel cells, solar cells, batteries, and supercapacitors. An electrochemical cell is basically comprised of an electrolyte and a pair of separated electrodes. Two electrodes facilitate the flow of electric current in a cell primarily by allowing ion diffusion to occur in the permeable electrolyte. An electrode serves as an electrical conductor that establishes contact with a non-metallic electrolyte in an electrical circuit. Electrode can be either cathode or anode which involves the reduction–oxidation (redox) reaction. Anode is the electrode where the electron is lost through an oxidation reaction, whereas a cathode is the electrode where the electron is gained through the reduction reaction. On the other hand, electrolyte also plays an important role in the electrochemical devices. Electrolyte is a medium to produce the ionized species or more precisely known as charge carriers. The mobile ions produced from the electrolyte will be transported in the electrolyte from an electrode to the opposite electrode and allowed to perform chemical reaction at the electrodes as long as there is an electric current applied across the cell. As a result, energy is delivered to the external circuit. Electrolytes can be divided into two main classes that are liquid electrolytes and solid-state electrolytes. In this chapter, the transition of electrolytes from liquid type to solid type is examined.

Electrolytes find broad usage in technology, particularly in electrochemical devices, spanning from the manufacture of commercial secondary rechargeable lithium-ion batteries to sophisticated high-energy electrochemical systems like chemical sensors, fuel cells, electrochromic windows (ECWs), solid-state reference electrode systems, memory devices, supercapacitors, and solar cells (Gray, 1991; Rajendran et al., 2004). The diverse applications of electrochemical devices include powering portable electronic devices such as laptops, mobile phones, and music players, as well as serving as the power source for hybrid electric vehicles (EVs) and start-light-ignition (SLI) systems used in traction (Gray, 1997; Ahmad et al., 2005).

DOI: 10.1201/9781003314424-5

5.2 DEVELOPMENT OF ELECTROLYTES

The liquid-based electrolyte faces several problems, especially the leakages of corrosive liquids and harmful gases. Therefore, researchers developed a new type of electrolyte, so-called solid-state electrolyte to overcome those shortcomings.

5.2.1 LIQUID ELECTROLYTES

Liquid electrolyte is a substance that conducts electricity through the migration of both cations and anions to the opposite electrodes when an electrical potential is applied. The doping salt is normally dissolved in an aqueous solution to produce an electrolyte solution. The electrolyte is then dissociated into cation and anion, and hence followed by a solvation process. Solvation is a process of association between molecules of solvent and molecules or ions of the salt via intermolecular interactions. In other words, the molecules or ions of the solute will be dissociated and then bounded by the molecules of solvent in the electrolyte solution. Hence, the surrounded ions will be separated and migrate to the opposite electrodes to generate the electric current when an electrical potential is applied to the electrolyte. Sodium chloride ($NaCl$) is a strong electrolyte because it achieves complete dissociation when it is dissolved in a solution due to the thermodynamic interactions between solvent and solute molecules. In contrast, acetic acid is a weak electrolyte as it cannot be dissociated completely.

Despite their widespread use, traditional liquid electrolytes suffer from several drawbacks including the corrosive nature of the solvent and its tendency of harmful gas leakage, electrolyte degradation, growth of lithium dendrites, unfavorable long-term stability as a result of liquid evaporation, and low safety feature arising from the presence of flammable organic solvents (Ramesh et al., 2011; Yang et al., 2008). These electrolytes also have a limited operating temperature range, are challenging to handle and manufacture due to the liquid phase, and have a lesser shelf life with a likelihood of experiencing internal short circuits (Gray, 1997; Stephan and Nahm, 2006). As a result, researchers have proposed the adoption of solid-state electrolytes as a replacement for traditional liquid electrolytes (Famprikis et al., 2019).

5.2.2 SOLID ELECTROLYTES

Solid electrolytes can be defined as non-aqueous materials that exhibit high conductivity due to the rapid diffusion of one ionic species through a lattice formed by immobile counterions. The conduction of solid electrolytes arises from the mobile ions, not electrons as the solid electrolytes have negligible electrical conductivity. Solid electrolytes are also named fast ion conductors. Some literature referred them as superionic conductors and optimized ionic conductors (Takahashi, 1989). Solid electrolytes are typically classified into four categories: crystalline solid electrolytes, glassy electrolytes, molten electrolytes, and polymer electrolytes.

In general, crystalline solid electrolytes are referred to those that utilize multiple divalent and/or trivalent cations based on ionic hopping within the crystal structures. It is noted that conduction pathways are interspersed throughout the immobile ion sub-lattice of the material. The mobile ions move along these channels from site to another site. Common examples of this type of electrolyte include β-alumina, γ-Li$_2$ZnGeO$_4$, and RbAg$_4$I$_5$ (Gray, 1997).

On the other hand, glassy electrolytes are categorized as amorphous solid conductors that are formed when ion-containing liquids are rapidly cooled below their glass transition temperature (T_g). In such event, the crystallization process is avoided. These vitreous materials consist of three main components: network formers, network modifiers, and ionic salts. The conductors in these glassy materials are either pure cation or pure anion (Gray, 1997). Glassy electrolytes exhibit some advantages over crystalline electrolytes because of their unique glassy state. These advantageous features are isotropic conduction, elimination of grain boundaries, high ionic conductivity with extremely low electronic contribution, and ease of preparation (Fusco and Tuller, 1989). These conventional glassy conductors remain in a glassy state at ambient temperature with T_g above room temperature. The mobile ions are transported between different fixed vacant sites in a frozen disordered structure of glassy electrolytes to generate ionic conductivity (Karan et al., 2008). Examples of glass electrolytes are Li$_2$S-P$_2$S$_5$, Li$_2$S-B$_2$S$_3$, Li$_2$S-SiS$_2$, and ortho-thiophosphate (Na$_3$PS$_4$) (Hayashi et al., 2012; Deshpande 2009).

Molten electrolytes consist of molten single salt or molten eutectic mixtures and exhibit high ionic conductivity, usually above 1 S/cm. Some examples of commonly used molten electrolytes include LiCl-KCl eutectic mixtures and chloroaluminates (AlCl$_3$-MCl, where M is any alkali metal). Despite that, the use of molten electrolytes requires meeting specific technical requirements to prevent cell corrosion and leakage (Gray, 1991). Although crystalline and glassy electrolytes have high ionic conductivity feature, they are brittle. The remedy for this drawback is the application of polymeric materials. With the addition of polymeric materials, the mechanical properties of the electrolyte can be enhanced. Such enhancement is partly due to the ability of the polymer materials to accommodate volume changes. As a result, solid polymer electrolytes (SPEs) have become a viable option for electrochemical applications. SPEs are frequently used along with intercalation materials like electrodes in rechargeable lithium-type batteries. SPEs differ from glassy conductors in terms of T_g. Most of the SPEs have T_g below ambient temperature. A different ion hopping mechanism is observed for SPEs in comparison to glassy electrolytes, as mentioned by Karan and co-workers (Karan et al., 2008).

5.2.3 SOLID-STATE POLYMER ELECTROLYTES

After World War II, the industrialized world experienced an economic downturn (Noto et al., 2011). The oil crisis in year 1973 remarked the end of the availability of cheap and abundant energy. Renewable energy sources are the breakthroughs to create new innovative devices for storage and energy conversion including new materials such as polymer electrolytes (Noto et al., 2011). In 1975, Wright, a polymer chemist from Sheffield, first developed conductive solid-state polymer electrolytes to address the limitations of traditional liquid electrolytes (Wright, 1975). SPEs have received an upsurge of interest in the electrochemical devices application since the year 1975 to improve the electrochemical performances of the

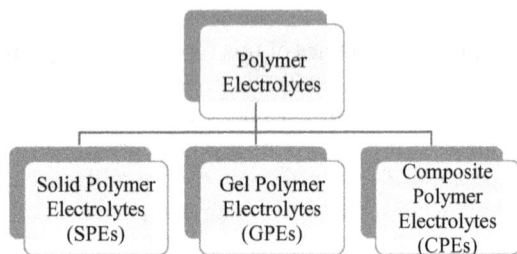

FIGURE 5.1 Sub-classes of polymer electrolytes.

devices. Solid-state polymer electrolytes can be categorized into several classes based on their composition and properties (Figure 5.1).

Solid-state polymer electrolyte plays three important roles in electrochemical devices. Polymer electrolyte serves as an electrode separator by insulating the opposite electrodes from each other. So, the addition of an inert porous spacer between the electrolyte and electrode interface is not required in this case. Polymer electrolyte plays an important role during the charging and discharging processes. It acts as a medium channel to generate ionic conductivity between the electrodes during the processes. Apart from that, it also works as a binder to ensure good interfacial contact with electrodes (Gray, 1991; Kang, 2004).

5.2.3.1 Solid Polymer Electrolytes

SPEs have become increasingly important in various applications, including energy storage and electrochemical displays, due to their flexible ion transportation properties (Wright, 1998). SPEs have been developed to replace conventional liquid electrolytes and offer enhanced safety due to their solvent-free condition. The use of SPEs in lithium-ion batteries has proven to be compatible with lithium metal electrodes, resulting in low self-discharge and closed-circuit discharge. Furthermore, SPEs are highly elastic and easy to handle, allowing for the creation of all solid-state electrochemical cells (Ibrahim et al., 2012). Compared to crystalline or glassy electrolyte-based cells, electrochemical cells based on SPEs exhibit commendable interfacial contact between the electrode and electrolyte, which can be maintained during charging and discharging processes (Gray, 1991, 1997). Additionally, SPEs do not generate internal pressure, which could lead to explosions during charge and discharge processes in electrochemical cells (Liew et al., 2013). According to Gray, the ionic transportation in polymer electrolytes relies on the local relaxation processes that occur within the polymer chains. These relaxation processes can provide similar properties to those of liquid electrolytes (Gray, 1991).

SPEs are unique systems that do not require solvents and instead rely on dissolving low-lattice-energy metal salts in the ion-coordinating molecules with aprotic solvents to create an ionically conducting pathway (Bruce and Vincent, 1993). The solvating group, or donor atom, of the polymer can form a covalent bond with the salt's cations for ion transportation. This coordination occurs through electrostatic interactions between the charges on the cation and the solvating group. Such dissociation of the coordination gives rise to ionic conduction in the polymer electrolytes. Armand's work has demonstrated new possibilities for solid-state ionic conduction, particularly in lithium battery fabrication (Armand, 1986). In his work, electrodes consisting of graphite intercalation compounds were coupled with polymer-salt complexes electrolytes. In addition to high safety performance, SPEs possess numerous advantages ranging from low requirement of vapor pressure to low volatility to high energy density, and other excellent properties (Gray, 1991, 1997; Adebahr et al., 2003; Nicotera et al., 2002; Ramesh and Liew, 2013; Stephan and Nahm, 2006).

SPEs possess numerous superior features, including inherent viscoelasticity, which is a mechanical behavior demonstrating both viscous and elastic properties. Besides having good mechanical properties, SPEs also have the ability to suppress the growth of lithium dendrite. Apart from that, they are lightweight, low in cost, and easy to handle and produce. One of the notable advantages of SPEs is their wide operating temperature range (Rajendran et al., 2004; Ramesh et al., 2011; Baskaran et al., 2007; Imrie and Ingram, 2000). Additionally, SPEs can be configured into various shapes, thanks to their highly flexible polymer membrane structure (Gray, 1991).

The first SPE generation which took in the form of crystalline poly(ethylene oxide) (PEO)–based polymer electrolytes was introduced by Wright's group (Fenton et al., 1973; Quartarone et al., 1998). Initially, they conducted research on the effect of alkali metal salts on PEO (Noto et al., 2011; Wright, 1975; Fenton et al., 1973). Given the good solvating properties exhibited by PEO, its electrolyte form has low ionic conductivity, falling in the range of 10^{-8}–10^{-7} S/cm. Such low value is attributed to its high crystallinity and tendency to recrystallize. Since then, the second generation of SPEs was developed to decrease the crystallinity of polymer electrolytes as the first generation of SPEs reached the same conclusions about the link between the amorphous region and the ionic conductivity. Various techniques have been employed to hinder polymer complexes from recrystallizing as well as to reduce the crystallinity degree. Examples of such techniques include modifying the polymer, blending polymers, and using semi-crystalline or amorphous polymers such as poly(methyl methacrylate) (PMMA). Other techniques include the incorporation of additives such as plasticizers, ionic liquids, and inorganic fillers.

One notable method of minimizing crystallization is structural modification. In this method, cross-linked is induced within the short chains of the ethylene oxide (EO) of the PEO (Quartarone et al., 1998). In one study, poly(dimethylsiloxane) (PDMS) was used to crosslink with PEO using an aliphatic isocyanate, resulting in copolymers with an ionic conductivity of approximately 10^{-5} S/cm after the inclusion of lithium perchlorate ($LiClO_4$) (Bouridah et al., 1985). Another study involved cross-linking PEO with poly(propylene oxide) (PPO), resulting in a LiClO4-doped polymer network. The produced polymer complex exhibited five folds of ionic conductivity compared to that of unlinked PEO polymer electrolytes (Watanabe et al., 1986). Additionally, a polyacrylonitrile-polyethylene oxide (PAN-PEO) copolymer was synthesized by Yuan et al., which showed higher ionic conductivity comparatively. The highest ionic conductivity of polymer electrolytes based on this copolymer and $LiClO_4$ (EO)/(Li) ratio of about 10 was recorded at 6.79×10^{-4} S/cm under ambient temperature (Yuan et al., 2005). Despite their good safety feature and high mechanical strength, SPEs still exhibit low ionic conductivity which hinders their expansion into electrochemical devices (Yang and Wu, 2022).

5.2.3.2　Gel Polymer Electrolytes

The third generation of gel polymer electrolytes (GPEs) was developed in the 1990s to address the low ionic conductivity of SPEs (Noto et al., 2011). GPEs are also known as gel ionic SPEs or plasticized PE. GPEs are formed via dissolution of polymer host with doping salt in organic solvents or plasticizers (Ramesh et al., 2012; Osinska et al., 2009; Rajendran et al., 2008). This immobilizes the liquid electrolyte in the polymer matrix, which gives GPEs its unique property (Han et al., 2002). GPEs possess both the cohesive properties of solids and the diffusive properties of liquids, with local relaxations that provide a liquid-like degree of freedom (Ramesh et al., 2012). Common components of GPEs include carbonate esters such as propylene carbonate (PC) or ethylene carbonate (EC), in high dielectric constant solvents like N,N-dimethylformamide (DMF) and γ-butyrolactone (Ramesh et al., 2011).

GPEs have a variety of inherent characteristics, including low interfacial resistance at the electrode–electrolyte boundary and low electrode materials reactivity. They also have better safety features along with shape flexibility. Additionally, they exhibit good ionic conductivity improvements with just little plasticizers (Ahmad et al., 2008; Pandey and Hashmi, 2009; Aruchamy et al., 2023). Compared to liquid electrolytes, GPEs offer improved electrochemical properties which include extending operating temperature conditions (Ahmad et al., 2005; Stephan et al., 2002). Besides being light in weight, GPEs are leakproof and easy to fabricate into any shape and size (Zhang et al., 2011). To improve GPE performance, plasticizers are commonly used, as seen in PEO-lithium trifluoromethanesulfonate (LiCF$_3$SO$_3$) electrolytes, where the addition of dibutyl phthalate (DBP) increases ionic conductivity about three orders from 7.13×10^{-7} S/cm to 6.03×10^{-4} S/cm (Sukeshini et al., 1998). Plasticizers can also increase the cationic transport number in the electrolyte system as seen in PEO-sodium metaphosphate (NaPO$_3$) added with PEG (Sukeshini et al., 1998).

However, GPEs also have some drawbacks. These include lower mechanical strength and poor dimensional stability. The slow evaporation of organic solvents during the fabrication process reduces the productivity of the GPE. Besides, they have reduced thermal, electrical, and electrochemical stabilities (Ramesh et al., 2011; Kim et al., 2006; Raghavan et al., 2010). GPEs are less compatible with lithium metal anode in lithium batteries because of their poor interfacial stability (Pandey and Hashmi, 2009). GPEs also exhibit poor electrochemical performances due to their narrow working potential window range and high vapor pressure (Ramesh et al., 2011; Kim et al., 2006; Raghavan et al., 2010). The shortcomings of traditional GPEs have led to the development of the next generation of polymer electrolytes based on composite.

5.2.3.3　Composite Polymer Electrolytes

Recently, researchers have shown a lot of interest in the fourth generation of SPEs, known as composite polymer electrolytes (CPEs). These polymer electrolytes are created by blending small amounts of organic or inorganic fillers with polymer electrolytes, resulting in what are referred to as nanocomposite polymer electrolytes (NCPEs) (Osinska, et al., 2009; Tang et al., 2021). By adding these fillers into GPEs, the physical and mechanical properties of polymer electrolytes are improved. CPEs offer many benefits, including good interfacial contact in the electrode–electrolyte region, high flexibility, enhanced ion transport, high ionic conductivity, and excellent thermodynamic stability toward lithium and other alkali metals (Gray, 1997). Additionally, they exhibit superior interfacial and electrochemical properties when used with a lithium metal anode (Stephan and Nahm, 2006). The capacity of lithium batteries can be increased by incorporating fillers, as shown in a study by Stephan and Nahm in 2006. CPEs are a promising option because of their excellent chemical, mechanical, thermal, and electrochemical stability (Noto et al., 2011). Some commonly used fillers include manganese oxide (MnO_2), titania (TiO_2), and alumina (Al_2O_3).

In the past, SiO_2 has been commonly used as a filler, and it has been found to significantly improve both the ionic conductivity and mechanical properties of PEO-based electrolytes. Fan et al. (2003) compared unmodified SiO_2 with silane-modified SiO_2 and found that the latter resulted in much higher ionic conductivity. Besides that, treatment of conventional fillers can inhibit recrystallization of PEO-based polymer electrolytes. In a study by Xi et al. (2005), solid acid sulfated-zirconia (SZ) was used as a filler in NCPEs based on PEO-LiClO$_4$, resulting in a two-order-of-magnitude increase in ionic conductivity compared to the pristine PEO-LiClO$_4$ polymer electrolytes. Furthermore, the ionic conductivity of polymer electrolytes with SZ was twice as high as untreated NCPEs (Xi et al., 2005). Nanosized organic–inorganic hybrid materials like zinc aluminate ($ZnAl_2O_4$) that have high surface area have also been introduced. When added to PEO-LiClO$_4$, 8 wt.% of $ZnAl_2O_4$ achieved higher ionic conductivity of 2.23×10^{-6} S/cm at ambient temperature (Wang et al., 2009). Such improvement was attributed to the reduction of crystallinity of the polymer membrane and the increase of lithium-ion transference number.

5.2.4   GENERAL REQUIREMENTS OF SOLID-STATE POLYMER ELECTROLYTES FOR ELECTROCHEMICAL DEVICES APPLICATIONS

Each electrochemical device application requires different basic requirements of solid-state polymer electrolytes. Solid-state polymer electrolytes with strong adhesive and superior chemical stability properties are highly recommended for all electrochemical device applications because these properties can reduce the interfacial contact between electrodes and the electrolyte. For a practical fuel cell application, proton-conducting polymer electrolytes are required. The protons will travel from the anode to the cathode to generate the electrical current in the fuel cell. In contrast, the main requirement of solar cells, particularly dye-sensitized solar cells, is the iodide electrolyte (Sagaidak et al., 2016). The iodide electrolyte will receive an electron from a positive electrode (or known as the cathode). The iodide (I^-)/triiodide (I_3^-) redox couple is regenerated on the cathode and thus the I_3^- is reduced to I^- in the mechanism. GPEs are commonly used as electrolytes in solar cell applications. On the contrary, ECWs application needs a transparent electrolyte film to observe the color changes on the electrolyte when voltage, light, or heat is applied. There are a few requirements that must take note for batteries, especially lithium batteries.

Solid-state polymer electrolytes must be free of water and inert to the reactive electrodes (the lithium electrodes for lithium batteries) for battery application to avoid explosion. The polymer electrolytes must have a wide electrochemical potential window (>3 V) which is the voltage range of a substance before undergoing decomposition. Unlike batteries, fuel cells, or solar cells, supercapacitors do not require any specific requirement. For supercapacitor applications, the charge carriers can be proton, lithium, sodium, magnesium, or potassium.

5.3 DESCRIPTION OF IONIC CONDUCTION

5.3.1 MECHANISM OF IONIC CONDUCTION

In general, diffusion involves the movement of atoms, molecules, or ions within a solid phase (Raghavan, 2004). This motion allows for the attainment of uniform concentrations of the diffusing species between different locations (Kudo and Fueki, 1990). In an electrolyte, the primary means of diffusion is through the movement of mobile charge carriers when an electric field is present. When there is no electric field, the ions in the electrolyte undergo Brownian motion, but when a voltage is applied, they migrate in the direction of the electric field, a process referred to as ion conduction. However, if the ions are trapped within the crystalline phase of a solid matrix, which is an ordered arrangement of molecules, they must overcome a potential barrier in order to continue the diffusion process (West, 1999). The ion transport is restricted in the crystalline region. Although ions in a lattice structure vibrate constantly, they usually lack sufficient thermal energy to break free from their lattice sites. However, if the ions can surpass the barrier, they can move to neighboring lattice sites, leading to ionic conduction, migration, hopping, or diffusion. In polymer electrolytes, ion diffusion can occur via two primary mechanisms: the vacancy mechanism and the interstitial mechanism. Both mechanisms are graphically represented in Figures 5.2–5.4.

Vacancy mechanism is defined as the hopping mechanism of an ion from its normal position to an adjacent, but empty, site (Souquet et al., 2010). The ions require sufficient energy which arises from the thermal energy of ionic vibrations

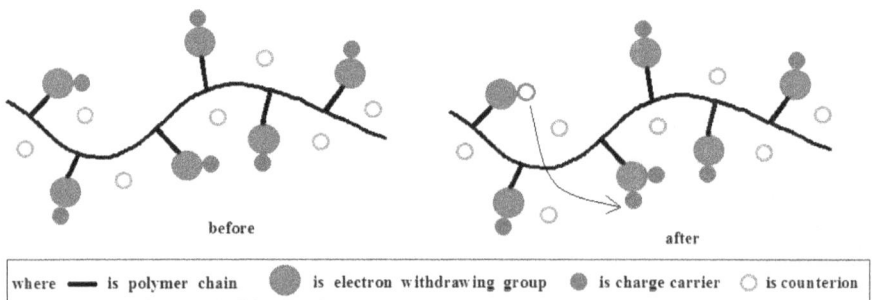

where ━━ is polymer chain   ● is electron withdrawing group   ● is charge carrier   ○ is counterion

FIGURE 5.2 Illustration of ion diffusion before and after a vacancy mechanism. Adapted from Souquet et al. (2010).

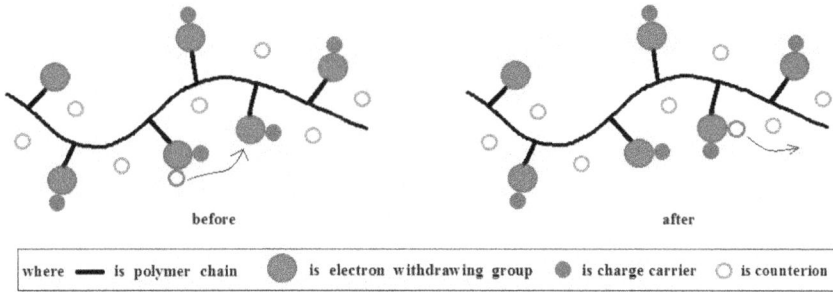

FIGURE 5.3 Graphical representation of ion diffusion of interstitial mechanism (Han et al., 2002).

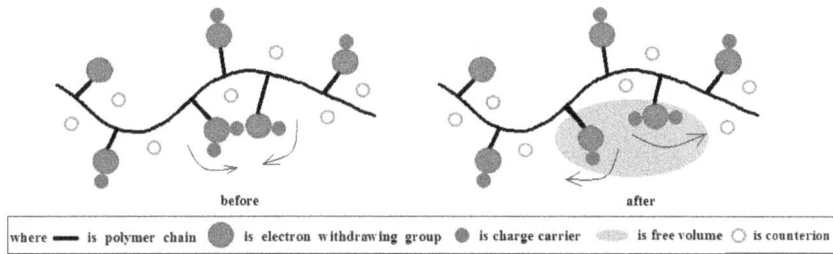

FIGURE 5.4 Graphical representation of ion diffusion of free volume mechanism coupled with the chain movement. Adapted from Souquet et al. (2010).

to break the coordination bonds and jump from a site to an adjacent empty site (Raghavan, 2004). On the other hand, the counterion from the salt can move freely in the electrolyte.

On the contrary, the interstitial mechanism involves the migration of a mobile charge carrier from one interstitial position to another, as represented in Figure 5.3 (Raghavan, 2004; Souquet et al., 2010; Smart and Moore, 2005). Interstitial diffusion rate is faster than vacancy diffusion. This is because the interstitials have weak bonding to the surrounding ions and there is a likelihood of empty adjacent interstitial site for the ions to move through during ion conduction (West, 1999).

Additionally, for polymer electrolytes in the above T_g condition, there is another mechanism of interstitial pair migration called the free volume mechanism, which is a cooperative process involving both interstitial pair migration and polymer segment mobility (Figure 5.4). During the local movement of polymer segments, a cage is created by the polymer's closest neighbors. This cage is subject to random density fluctuations or chain movements, which generate free volume within the polymer matrix. Subsequently, mobile species have the ability to move out of their original cage and into another when the fluctuation in polymer chain density creates an adjacent cage that is spacious enough for ion transport (Souquet et al., 2010).

5.3.2 Requirements of Ionic Conductivity

The ionic hopping process is influenced by the movement of mobile charge carriers, or ions, that have been dissociated from the polymer complexes. To enable this conduction mechanism in polymer electrolytes, five fundamental requirements must be met. First, there must be mobile ions that can migrate. Then, there should be vacant sites available for the ionic hopping process to occur, as ions will be mobile if there is a vacancy availability for the ions to occupy. Next, the potential energies of the empty and occupied sites should be similar, and there should be a low activation barrier for ions to jump from one site to the next. This is important because many vacant sites are of no use if the ions cannot enter due to small vacant sizes. Other than that, the structure should have a certain degree of framework that allows open pathways for mobile ions to diffuse. Finally, the framework should be highly polarizable as well (West, 1999).

5.3.3 Governing Parameters of Ionic Conduction

The ionic conductivity of a polymer electrolyte is expressed as follows:

$$\sigma(T) = \sum_i n_i q_i \mu_i$$

where n_i is the number of charge carriers of type i per unit volume, q_i is the charge of ions of type i, and μ_i is the ionic mobility of type i which is a measure of the drift velocity in a constant electric field (Gray, 1991, 1997; Smart and Moore, 2005). The amount of free charge carriers and their mobility in polymer electrolytes are key factors that strongly influence the ionic conductivity of such materials. To be considered conductive, polymer electrolytes must meet certain criteria, including the ability to detach more mobile charge carriers from coordination bonds and transport these charge carriers with high mobility. There are several factors that can affect these criteria, such as the degree of crystallinity and the dielectric constant of the polymer. In general, a low degree of crystallinity (or more amorphous) with a low T_g value is preferred. Aside from that, high ion mobility with high concentration of mobile ions is also important. In a polymer electrolyte system, crystallinity is referred to as a physical state in which solid materials have ordered and aligned arrangements of atoms, molecules, or ions, while amorphous refers to a state where these arrangements are random and disordered (Steven, 1999).

Pure crystalline polymer electrolytes are challenging for ion transportation because their molecules are tightly packed, making it difficult for ions to break coordinative bonds between cations from salt and electron-withdrawing groups from macromolecules. Therefore, the presence of crystallographic defects in polymer electrolytes can facilitate ionic conduction by creating amorphous regions where charge carriers can diffuse efficiently. When the polymer electrolyte reaches its glass transition temperature, T_g, it transitions from a glassy to a rubbery state in the amorphous region. Despite the short-range vibrations and rotations that occur below T_g, ion mobility remains restricted in the glassy state. Such restriction hinders the

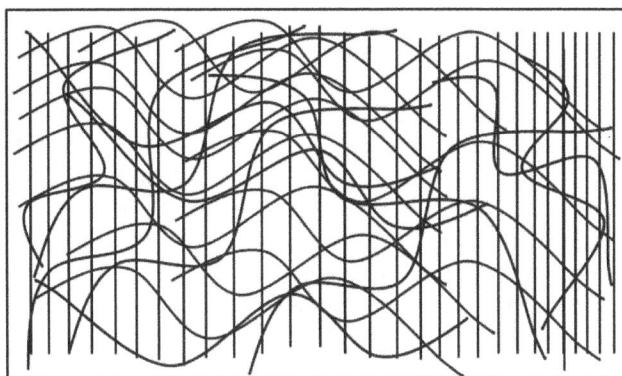

where

		Crystalline region
$\wedge\!\wedge$	Amorphous region

FIGURE 5.5 Illustration of mixed amorphous and crystalline regions in semi-crystalline polymer structure.

ion diffusion in polymer electrolytes. Above T_g, the rubbery state allows for long-range molecular motion, promoting segmental movement among the molecules or ions in the polymer chains. The electrostatic forces between the ions can be reduced by using polymers having high dielectric, which promotes increasing the distance between them and preventing ion aggregation and clustering (Eliasson et al., 2000). In a nutshell, polymer electrolytes with high dielectric constant stimulate ion diffusion and enhance ion hopping. The number density and mobility of charge carriers are largely affected by the flexibility of the polymer chains. This is because solvating ions can readily dissociate from the interactive bonds in polymer chains, thereby increasing ionic conduction in polymer electrolytes.

5.3.4 Methods of Enhancing Ionic Conduction Mechanism

The limited applicability of polymer electrolytes is attributed to the low ionic conductivity. To increase the ionic conductivity to approximately mS/cm, several methods have been employed, including blending different polymers, modifying polymers like irradiating with gamma rays or using mixed salt systems, and adding various additives (Li et al., 2023). Examples of commonly used additives are plasticizers, ionic liquids, and fillers.

5.3.4.1 Modifications of Polymer

The conduction route in electrolytes can be affected by the degree of polymer crystallinity. Because of this reason, researchers have made alterations to the polymer to enhance its amorphousness. To achieve this, they focused on developing various macromolecular architectures, including various forms of copolymers (block and comb copolymers, etc.) and network polymers.

A block copolymer is one type of copolymers made up of blocks of different polymerized monomers where two homopolymers joined together at the ends. Comb copolymer and graft copolymers are segmented copolymers with a linear backbone of one group of polymerized monomers and covalently bonded branches of another group. The difference between comb copolymers and graft copolymers arises from the branching of the polymers. Comb copolymers have two or more different branches, whereas graft copolymers consist of one or more side chains which are different, structurally or configurationally from the main chain.

Network copolymers are branched copolymers in which all the polymer chains are interconnected to form a network by crosslink. According to Fish et al., polymer-salt complexes that exhibit chain flexibility and lower glass transition temperatures could enhance polymer segmental mobility. This in turn promotes ion diffusion through the polymer matrix (Fish et al., 1988). Their study demonstrated that when poly(methylsiloxane) complexes with oligo(oxyethylene) side chains are bound to $LiClO_4$, an ionic conductivity value as high as 7×10^{-5} S/cm was observed. At higher temperature ranges, the conductivity exceeded 10^{-4} S/cm (Fish et al., 1988).

In a separate study, Soo et al. synthesized and electrochemically characterized block copolymer electrolytes based on poly(lauryl methacrylate)-b-poly(oligo (oxyethylene) methacrylate) in the rubbery condition (55). These resultant electrolytes had better ionic conductivity and dimensional stability. Compared to the glassy-type block copolymer systems, potential window could be enhanced up to 5 V. When this polymer was applied in the battery, the device showed high reversible capacity and excellent capacity retention (Soo et al., 1999).

Guilherme et al. developed a block copolymer electrolyte consisting of polyethylene-b-poly(ethylene oxide) (PE-b-PEO) and $LiClO_4$. The highest ionic conductivity of this polymer electrolyte was achieved in the sample with the addition of 15 wt.% $LiClO_4$, reaching 3×10^{-5} S/cm as measured at ambient temperature (Guilherme et al., 2007). At 100°C, the ionic conductivity increased to $\sim 10^{-3}$ S/cm, but the degree of crystallinity of the polymer electrolytes was reduced due to block copolymerization (Guilherme et al., 2007).

Additionally, Lyons et al. (1996) synthesized polysilane comb polymers, $((CH_3CH_2OCH_2CH_2O(CH_2)_4)Si(CH_3))_n$, with ethoxyethoxybutane incorporated in the polymer's side chain, which served as a host polymer for the preparation of polymer electrolytes. The polymer electrolytes based on this comb polymer host system exhibited ionic conductivity of 1.2×10^{-7} S/cm at (Li)/(O) = 0.25 under ambient temperature. On the other hand, when the comb-branch copolymers were synthesized by copolymerization between poly(ethylene oxide methoxy) acrylate and lithium 1,1,2-trifluorobutane sulfonate acrylate, new fluorinated copolymers would be produced with enhanced ionic conductivity and lower T_g value (Cowie and Spence, 1999).

Polyethers with comb-shaped structures have been synthesized using poly(4-hydroxystyrene) (PHSt) as a multifunctional initiator and graft polymerization of EO or a mixture of EO and propylene oxide (PO). These polyethers, when combined with lithium triflate ($LiCF_3SO_3$), showed an ionic conductivity of approximately 10^{-5} S/cm at room temperature. The grafting process greatly reduced the crystallinity of these polymer electrolytes (Jannasch, 2000). In a separate study, Ding et al. used a combination of stable free radical polymerization (SFRP) and emulsion polymerization to synthesize graft copolymers comprising graft chains of macromonomer poly(sodium styrenesulfonate) (macPSSNa) and polystyrene (PS) backbone (Ding et al., 2002). These graft polymer electrolytes showed better proton conductivity compared to samples synthesized from random copolymers styrenesulfonic acid and styrene (PS–rPSSA), despite lower water uptake (Ding et al., 2002).

Last but not least, a network polymer based on poly(2-(2-methoxyethoxy)ethyl glycidyl ether) (PME2GE) has been used as a host polymer in random copolymer electrolytes containing $LiClO_4$ as the dopant salt. This system achieved a maximum ionic conductivity of approximately 10^{-4} S/cm at 40°C under the optimal composition of EO/ME2GE = 70/30 (Kono et al., 1993).

Polymer membranes with high ionic conductivity were synthesized using network polymers with hyperbranched ether side chains. To achieve this, a monosubstituted-epoxide monomer called 2-(2-methoxyethoxy)ethyl glycidyl ether (MEEGE) was copolymerized with EO. The copolymerization method employed the base-catalyzed anionic ring-opening polymerization with 2-(2-methoxyethoxy) ethanol. The synthesis resulted in the formation of semiterechelic poly(ethylene oxide-co-2-(2-methoxyethoxy)ethyl glycidyl ether) (P(EO/MEEGE)) oligomers, which were esterified with acrylic acid to produce polyether macromonomers. Network polymer electrolytes were prepared via photo cross-linking a mixture of the polyether macromonomer, lithium bis(trifluoromethylsulfonyl)imide (LiTFSI) salt, and a photo initiator. These electrolytes displayed ionic conductivity values of 1×10^{-4} S/cm at 30°C and 1×10^{-3} S/cm at 80°C (Nishimoto et al., 1999). However, this polymer modification technique was not found to be particularly effective in significantly improving the ionic conductivity of polymer electrolytes.

5.3.4.2 Polymer Blending

Blending polymers is a viable method to improve the ionic conductivity of polymer electrolytes. This process entails physically mixing two or more distinct polymers or copolymers without any covalent bonding. Typically, one of the materials in the mixture absorbs the active transporting species, while the other material provides mechanical support to the electrolyte system. In some cases, the inertness feature is contributed. Polymer blending has several benefits, including simplified sample preparation and the ability to regulate the physical properties of the polymer membrane through compositional changes (Rajendran et al., 2002). Additionally, polymer blending is cost-effective and doesn't require a polymerization process, unlike polymer modifications. The properties of polymer blends largely depend on the properties of the participating polymers, both chemical and physical. Not to forget if the resulting blend is homogenous or heterogeneous. If a few distinct polymers are capable of dissolving successfully in a typical solvent, they will create a homogenous blend or intermixing of the dissolved polymers, thanks to the rapid establishment of thermodynamic equilibrium (Braun et al., 2005).

Several dual polymer electrolyte systems have been created and studied, including PMMA-PVC (Stephan et al., 2002; Choi and Park, 2001; Rajendran et al., 2000), PVA-PMMA (Rajendran et al., 2004), PMMA-poly(vinylidene fluoride) (PVdF) (Cui et al., 2008), poly(vinyl acetate) (PVAc)-PMMA (Baskaran et al., 2006b), PVAc-PVdF (Baskaran et al., 2006a), PEO-PVdF (Yang et al., 2008), PVC-poly(ethyl methacrylate)

(PEMA) (Rajendran et al., 2008; Han et al., 2002), and PVA-poly(styrene sulfonic acid) (PSA) (Kumar and Bhat, 2009). In the application of polymer electrolytes, combination of PMMA and poly(vinyl chloride) (PVC) is widely used for the polymer host. PMMA polymer is amorphous in nature with brittle characteristics. To overcome this weakness, PVC is introduced as a mechanical stiffener. PVC's characteristic is attributed to the dipole–dipole interaction between its hydrogen and lone pair electrons of the chlorine atom (Ramesh et al., 2013).

The combination of PMMA and PVC can be observed through the change in the shape of the particles from spherical to rectangular, as observed from SEM images in the work of Ramesh et al. (2010). A polymer electrolyte consists of 70 wt.% PMMA and 30 wt.% PVC polymer blends with 10 wt.% LiTFSI exhibited a room-temperature ionic conductivity of 1.60×10^{-8} S/cm. Beyond this ratio, phase separation occurred in the polymer electrolyte. Consequently, PVC's globular agglomeration as a result of high PVC loadings could obstruct ion transport in the polymer electrolyte system (Ramesh et al., 2010).

Ionic conductivity in polymer electrolytes was found to be higher in other polymer blends compared to PMMA-PVC blend systems. For instance, Sivakumar et al. observed that a PMMA–PVdF blend GPE with a PVdF:PEMA ratio of 90:10 produces 1.50×10^{-4} S/cm of ionic conductivity value (Sivakumar et al., 2007). However, the ionic conductivity value decreased at higher PEMA content due to increased crystallinity in PEMA domains. Nevertheless, this GPE displayed favorable transport properties and better interfacial stability with Li electrodes.

Wu et al. (2009) developed a promising LiTFSI-based polymer electrolyte by blending hyperbranched polyether with poly(3-{2-(2-(2-hydroxyethoxy) ethoxy) ethoxy}methyl-3'-methyloxetane) (PHEMO), and poly(vinylidene fluoride-hexafluoropropylene) (PVDF-HFP) as a host polymer. This novel polymer electrolyte exhibited a maximum ionic conductivity of 1.64×10^{-4} S/cm at 30°C and has a wide electrochemical potential window up to 4.2 V vs. Li+/Li and high decomposition temperature above 400°C, making it a suitable candidate for use as an electrolyte in lithium-ion batteries (Rahaman et al., 2014). Unfortunately, the ionic conductivity is still less than mS/cm, prompting research in exploring other methods that can significantly increase the ionic conductivity of polymer electrolytes, in place of the polymer blending method.

5.3.4.3 Gamma Irradiation

Utilizing gamma-ray radiation as a means of enhancing the ionic conductivity of polymer electrolytes is a viable approach. When polymeric materials are exposed to ionizing radiation, it can generate reactive intermediate products like excited states, ions, and free radicals (Rahaman et al., 2014). These free radicals, which are produced due to gamma irradiation, have the ability to impact the structure of the polymer chains by means of intermolecular cross-linking or chain scission (Nanda et al., 2010; Sinha et al., 2008). Additionally, gamma irradiation can modify various properties of the polymer complexes (Rahaman et al., 2014). By absorbing high energy, gamma irradiation can restrain the crystalline region, alter the molecular weight distribution, boost the ionic conductivity, or even enhance the mechanical strength of the polymer electrolytes (Nanda et al., 2010; Damle et al., 2008; Ghosal et al., 2013).

In Song et al. work, PEO was cross-linked with $LiClO_4$ using gamma radiation, and subsequently, PVdF was blended with cross-linked PEO to produce polymer blend electrolytes (Song et al., 1997). The blend was then exposed to gamma radiation to create a simultaneous interpenetrating network (SIN). It is noted that these SIN polymer electrolytes induced by gamma radiation not only showed a high mechanical modulus of 10^7 Pa but also a room-temperature ionic conductivity exceeding 10^{-4} S/cm.

Tarafdar et al. also conducted gamma radiation research, developing polymer electrolytes based on gamma-irradiated PEO, and observed that ionic conductivity was substantially improved at 35 kGy of gamma radiation dose, which is attributed to the decreased crystallinity of the polymer (Tarafdar et al., 2010). An SPE consisting of γ-irradiated PVdF and lithium bis(oxalato)borate (LiBOB) achieved a maximum ionic conductivity of 3.05×10^{-4} S/cm, which is 15% higher than the polymer electrolyte prepared without gamma radiation (Rahaman et al., 2014). While gamma radiation is a promising method for enhancing ionic conductivity, high doses of it may degrade the polymer electrolytes (Akiyama et al., 2010).

5.3.4.4 Mix Salt System

The ionic conductivity of polymer electrolytes can be improved by mixing dual salts, which can prevent the formation of aggregates and clusters, thereby increasing the mobility of ion carriers (Gray, 1997). Additionally, this mixed dual-salt system can deliver more mobile charge carriers for ion transport, resulting in higher ionic conductivity compared to a single-salt system. Arof and Ramesh developed polymer electrolytes based on a dual-salt system consisting of PVC and $LiCF_3SO_3$, with lithium tetrafluoroborate ($LiBF_4$) as doping salts. Although the ionic conductivity only increased slightly compared to that single-salt system, the enhancement is attributed to the improved mobility of charge carriers by avoiding the aggregation effect (Arof and Ramesh, 2000).

Yang et al. also prepared a mixed salt system of GPEs consisting of PVdF/poly((ethylene glycol) diacrylate) (PEDGA)/PMMA and a mixture of salts comprising of lithium hexafluorophosphate ($LiPF_6$)/$LiCF_3SO_3$ (Yang et al., 2006). The ionic conductivity of the mixed salt system polymer electrolytes was five times higher than that of the $LiCF_3SO_3$-based polymer electrolyte system. With polymer electrolytes having 10 wt.% of $LiPF_6$ and 1 wt.% of $LiCF_3SO_3$, ionic conductivity as high as 1.5 mS/cm was observed. This polymer electrolyte also had a stable electrochemical potential range. However, the limitation of this technique is the solubility of dual salts in the same solvent, which requires researchers to ensure that both salts can be solubilized in the same solvent.

5.3.4.5 Additives

To enhance the ionic conductivity, different kinds of additives can be employed, including plasticizers, fillers, and ionic liquids.

5.3.4.5.1 Plasticizers

Plasticization is commonly acknowledged as an efficient technique to significantly enhance the ionic conductivity of polymer electrolytes by reducing their degree of crystallinity (Suthanthiraraj et al., 2009). A plasticizer, which is a low molecular weight and non-volatile aprotic organic solvent with a glass transition temperature close to −50°C (Ramesh et al., 2012), is used in plasticized GPE. Plasticizers such as PC and EC as shown in Figure 5.6 are widely used (Suthanthiraraj et al., 2009; Ning et al., 2009; Pradhan et al., 2005).

The impact of plasticizers on polymer electrolytes is determined by their properties such as its viscosity, dielectric constant, interaction with the polymer, and coordinative bonding with the ions (Rajendran and Sivakumar, 2008).

In general, the presence of plasticizers in polymer electrolytes improves the solvating degree of the salt, boosts ion mobility, and improves the contact between the electrolyte and the electrode interfaces (Rajendran et al., 2004; Ramesh and Arof, 2001). Moreover, plasticizers are advantageous additives due to their high miscibility with polymers and low viscosity which enhance processability (Ramesh and Chao, 2011). Besides that, they usually have a high dielectric constant. Adding plasticizers to the polymer electrolyte system is a proven technique for improving ionic conductivity without affecting other properties of the polymer electrolytes (Ganesan et al., 2008). Understanding the roles of plasticizers in enhancing the ionic conductivity of polymer electrolytes is crucial. The addition of plasticizers is expected to increase the ionic conductivity of polymer electrolytes by modifying the local structure, increasing the amorphous segment, and altering the local electric field distribution of the polymer matrix.

The primary role of plasticizers is to lower the T_g of polymer electrolytes. The effect of this is the reduction of polymer's modulus at the defined temperature. An ideal solid-state polymer electrolyte must have high ionic conductivity and excellent

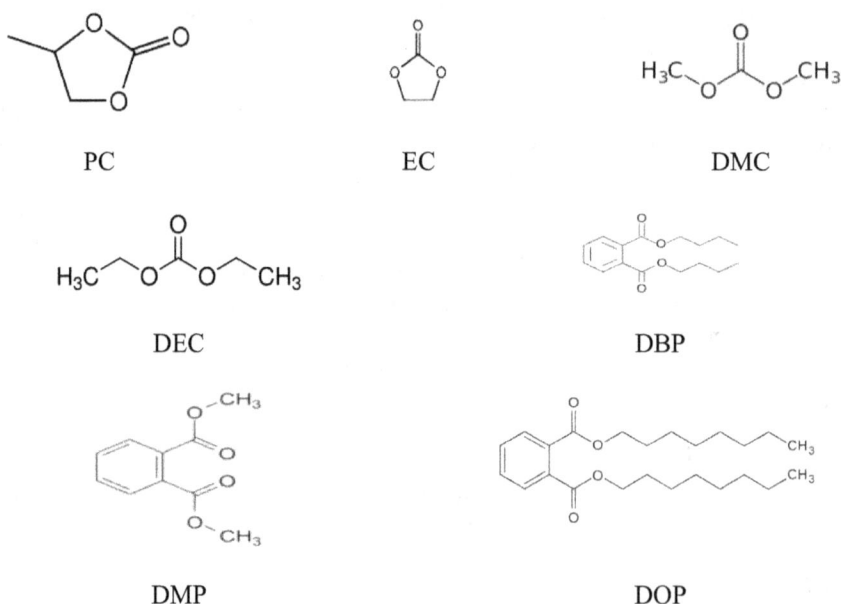

FIGURE 5.6 The chemical structure of some common plasticizers.

mechanical properties. However, the mechanical behavior of polymer electrolytes is reduced with mass loadings of plasticizers. Plasticizer undergoes a transition from a glassy state to a rubbery state at lower temperatures. In addition, plasticizers decrease the viscosity of the polymer system, facilitating the transport of ions within the polymer complexes. Besides, plasticizers can reduce the coordinative interactions within the polymer chains, improving their flexibility within the polymer matrix which favors ionic migration. This encourages the development of free volume within the polymer matrix, thereby enhancing the long-range segmental motion of the polymer system when it is swollen with a plasticizer (Ganesan et al., 2008).

To examine the effect of plasticizers on the ionic conductivity of the PEO-LiClO$_4$ polymer complex, three types of ester class plasticizers – dioctyl phthalate (DOP), DBP, and dimethyl phthalate (DMP) – were reviewed. Among them, DOP has better thermal stability, as demonstrated in differential thermal analysis. Michael's findings showed that the weight loss reduced as the plasticizer concentration increased (Michael et al., 1997).

In a study by Ali et al. (2007), plasticized-polymer electrolytes were synthesized from PMMA with PC or EC as the plasticizer, and LiTf or LiN(CF$_3$SO$_2$)$_2$ as the dopant salt. Their result showed that ionic conductivity increased with the concentration of the plasticizer. They also observed that PC plasticized-polymer electrolytes exhibited higher ionic conductivity compared to that of EC plasticized-polymer electrolytes.

Incorporating plasticizers in polymer electrolytes composed of PVA/PMMA-LiBF$_4$ was investigated by Rajendran et al. (2004). The highest ionic conductivity was observed at 1.29 mS/cm for the EC complex due to its high dielectric constant ($\varepsilon = 85.1$). Moreover, a maximum electrical conductivity of 2.60×10^{-4} S/cm at 300 K was achieved for electrolyte with 30 wt.% PEG as the plasticizer. Compared to the pure PEO-NaClO4 system, the ionic conductivity of the plasticized-polymer electrolyte increased by two orders of magnitude from 1.05×10^{-6} S/cm. This indicates that incorporation of plasticizer enhances the amorphous phase and, at the same time, reduces the energy barrier for ion transport, ultimately leading to higher mobility of lithium ions (Kuila et al., 2007). Despite its ability to extensively improve ionic conductivity, plasticization has some limitations. Among them are low safety performances and poor mechanical and thermal stabilities. Besides that, plasticization has a slow evaporation effect and produces a poor interfacial stability with lithium electrodes. In some cases, it also results in a narrow electrochemical window and low flash point (Ramesh et al., 2011; Pandey and Hashmi, 2009). Below is the summary of findings of some selected plasticized-polymer electrolyte systems (Table 5.1).

5.3.4.5.2   Ionic Liquids

Efforts have been made to address issues surrounding inadequate safety measures, subpar interfacial stability, and deficient electrochemical properties. To replace plasticizers, researchers have synthesized and created room-temperature ionic liquids (RTILs) in the past few years (Yim et al., 2015). Ionic liquids (ILs) are defined as nonvolatile molten salts which usually have melting temperature below 100°C (Pandey and Hashmi, 2009). They are also regarded as molten salts that remain liquid at room temperature (Quartarone and Mustarelli, 2011). ILs consist of a bulky and asymmetric

TABLE 5.1

The Ionic Conductivity of Selected Plasticized-Polymer Electrolyte System

Polymer Electrolyte System	Ionic Conductivity (S/cm)	References
PEO/PPO-LiCF$_3$SO$_3$-EC	3.00×10^{-4} S/cm at 303 K	Morales and Acosta (1997)
PEO/PPO-LiCF$_3$SO$_3$-PC	3.80×10^{-4} S/cm at 303 K	Morales and Acosta (1997)
PEO/PPO-LiCF$_3$SO$_3$-(γ-butyrolactone (γ-BLA))	1.53×10^{-6} S/cm at 303 K	Morales and Acosta (1997)
PEO/PPO-LiCF$_3$SO$_3$-ethyleneglycol dimethylether (EGDME)	3.40×10^{-6} S/cm at 303 K	Morales and Acosta (1997)
PEO/polyphosphazene (PPz)-LiCF$_3$SO$_3$-EC	5.94×10^{-4} S/cm at 303 K	Morales and Acosta (1997)
PEO/polyphosphazene (PPz)-LiCF$_3$SO$_3$-PC	5.30×10^{-4} S/cm at 303 K	Morales and Acosta (1997)
PEO/polyphosphazene (PPz)-LiCF$_3$SO$_3$-γ-BLA)	1.23×10^{-6} S/cm at 303 K	Morales and Acosta (1997)
PEO/polyphosphazene (PPz)-LiCF$_3$SO$_3$-EGDME	1.06×10^{-6} S/cm at 303 K	Morales and Acosta (1997)
PVC/PMMA-LiAsF$_6$-DBP	3.97×10^{-5} S/cm at 304 K	Rajendran et al. (2000)
P(VdF-co-HFP)/PVAc-LiClO$_4$-(EC+PC)	2.30×10^{-3} S/cm at 298 K	Choi et al. (2001)
PEO/P(VdF-HFP)-LiClO$_4$-(EC+PC)	1.30×10^{-4} S/cm at 303 K	Fan et al. (2002)
P(VdF-co-HFP)-LiClO$_4$-(EC+PC)	1.25×10^{-3} S/cm at 303 K	Fan et al. (2002)
Chitosan-lithium acetate (LiOAc)-oleic acid (OA)	1.05×10^{-5} S/cm at room temperature	Yahya and Arof (2003)
(PEO)$_8$-LiAsF$_6$-dibutyl sebacate (DBS)	$\sim 10^{-6}$ S/cm at room temperature	Ragavendran et al. (2004)
Poly(epichlorohydrin-co-ethylene oxide), P(EPI-EO)-NaI-I$_2$-poly(ethylene glycol) methyl ether (P(EGME))	1.70×10^{-4} S/cm at room temperature	Nogueira et al. (2006)
PAN-LiCF$_3$SO$_3$-EC	1.32×10^{-3} S/cm at room temperature	Ahmad et al. (2011)
PAN-LiCF$_3$SO$_3$-PC	8.64×10^{-4} S/cm at room temperature	Ahmad et al. (2011)
PMMA-NaClO$_4$-DMF	1.94×10^{-6} S/cm at room temperature	Sekhar et al. (2012)
PMMA-LiClO$_4$-(PC+DEC)	0.40×10^{-5} S/cm at room temperature	Sharma et al. (2012)
PMMA-LiClO$_4$-DBP	5.64×10^{-3} S/cm at 353 K	Kuo et al. (2013)
Poly (viny pyrrolidone) (PVP)-sulfamic acid (SA)-PEG	4.26×10^{-5} S/cm at 303 K	Bella et al. (2014)
Carboxy methylcellulose (CMC)-OA-glycerol	1.64×10^{-4} S/cm at 303 K	Chai and Isa (2014)

organic cation and an anion. A range of ionic liquids exist, with examples of organic cations include 1,3-dialkylimidazolium, N-methyl-N-alkylpyrrolidinium, tetraalkyl-ammonium, and guanidinium, to name a few (Jain et al., 2005; Ye et al., 2013). In contrast, common inorganic anions such as acetate (CH_3COO^-), nitrate (NO^{3-}), triflate (Tf$^-$), tetrafluoroborate (BF^{4-}), bis(trifluoromethylsulfonyl imide) (TFSI$^-$), bis(perfluoroethyl sulfonyl) imide (N(C$_2$F$_5$SO$_2$)$_2^{2-}$), hexafluorophosphate (PF^{6-}), and halides (Cl$^-$, Br$^-$, and I$^-$) have been widely adopted in ILs.

ILs exhibit unique properties that are influenced by the combination of cations and anions (Jain et al., 2005; Vioux et al., 2010). These properties make them a promising choice for use in electrolyte systems. Among their notable features are a wide electrochemical potential window of up to 6 V, a broad temperature range for decomposition, and their eco-friendliness as they are non-toxic and non-volatile (Ramesh et al., 2011; Cheng et al., 2007; Patel et al., 2011; Pandey and Hashmi, 2013). Furthermore, RTILs possess outstanding characteristics such as excellent chemical, thermal, and electrochemical stabilities, high ionic conductivity due to a high concentration of ions, good oxidative stability, superior ion mobility, and a high cohesive energy density. These properties make them ideal for use in polymer electrolytes (Bella et al., 2014). ILs also have exceptional solubility for a wide range of organic, inorganic, and organometallic compounds, as well as superior safety performance (Vioux et al., 2010; Reiter et al., 2006).

The strong plasticizing effect of ILs makes the polymer chains in the electrolyte system to be flexible via softening of the polymer backbone. This promotes the movement of mobile charge carriers in the polymer matrix. As a result, ionic conductivity is increased. On top of that, the low-viscosity nature of the ILs is responsible for the disruption of the orderly arrangement of the backbone. This gives rise to reduced crystalline regions of the polymer matrix (Singh et al., 2009). This creates more voids and free spaces for ion migration, enhancing the ionic mobility in the polymer system. And likewise for the ionic conductivity value.

When the bulky cations are paired with anions in ILs, packing efficiency tends to be low. This facilitates ion detachment from the ionic compound, resulting in better ionic conductivity of the polymer electrolytes. The incorporation of ionic liquids also produces sticky GPE. Such GPEs are beneficial for the design and application of electrochemical devices where better contact between the electrolyte and electrode interfaces is made possible (Reiter et al., 2006). The immobilization of ILs within the polymer matrices enables the application in the solid state, where drawbacks related to shaping and the risk of leakage are minimized. The incorporation of ionic liquid in the polymer electrolytes can produce a highly conducting ionic liquids-integrated polymer framework as shown in Figure 5.7 (Park et al., 2013). The mobile charge carriers are thus dissociated from the temporary coordination bond with the electron-withdrawing group of polymers. These charge carriers are eventually transported to the adjacent empty site.

In the past few years, many researchers have extensively studied the impact of incorporating ionic liquids into polymer electrolytes. For example, Sirisopanaporn et al. (2009) developed flexible, transparent, and freestanding GPEs by embedding N-n-butyl-N-ethylpyrrolidinium N,N-bis(trifluoromethane)sulfonimide-lithium N,N-bis(trifluoromethane) sulfonamide (Py$_{24}$TFSI-LiTFSI) ionic liquid solutions in PVdF-co-HFP copolymer matrices. The produced membranes exhibited good ionic conductivity ranging from 0.34 to 0.94 mS/cm at room temperature and could operate up to 110°C without physical degradation or leakage of IL within 4 months of storage time. In another study, a proton-conducting PVdF-co-HFP copolymer membrane containing

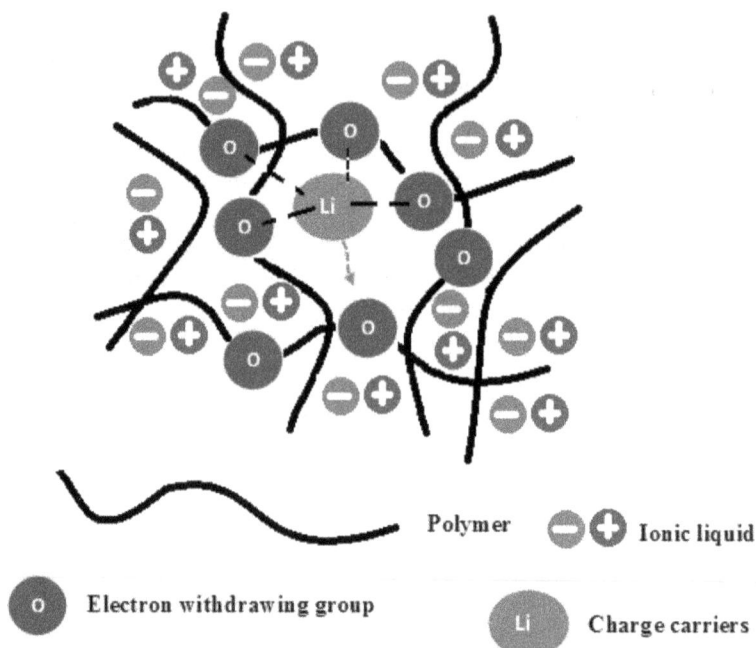

FIGURE 5.7 Schematic diagram of ion transport along the ionic liquid–integrated polymer framework. Adapted from Park et al. (2013).

2,3-dimethyl-1-octylimidazolium trifluromethanesulfonylimide (DMOImTFSI) was prepared. It showed good mechanical stability and achieved an ionic conductivity value of 2.74 mS/cm at 130°C (Sekhon et al., 2006).

IL is also incorporated into biodegradable polymers to form biopolymer electrolytes. For instance, IL plasticized-corn starch films were synthesized by Ning et al (Akiyama et al., 2010). The addition of 30 wt.% of 1-ally-3-methylimidazolium chloride (AmImCl) resulted in a maximum conductance of $10^{-1.6}$ S/cm. In our previous works, we prepared biopolymer electrolytes containing corn starch, $LiPF_6$, and ILs, namely 1-butyl-3-methylimidazolium hexafluorophosphate ($BmImPF_6$) or 1-butyl-3-methylimidazolium trifluoromethanesulfonate (BmImTf), using the solution casting technique (Ramesh et al., 2011; Liew and Ramesh, 2013, 2014). In both systems, the addition of ILs increased the ionic conductivity by three orders of magnitude, with the highest value observed at 1.47×10^{-4} S/cm for the sample with 50 wt.% of BmImTf (Singh et al., 2009). The Tf-based system exhibited even higher ionic conductivity, with a maximum of 3.21×10^{-4} S/cm (Liew and Ramesh, 2013, 2014). Doping ILs is an effective way to significantly enhance the ionic conductivity of polymer electrolytes without compromising their integrity.

5.3.4.5.3 Fillers and Nanofillers

Filler is a substance used to enhance the physical and mechanical characteristics of materials. When fillers are introduced into polymer electrolytes, they can create CPEs. NCPEs, on the other hand, are produced by dispersing nanometer-sized fillers into the polymer matrix. Nanotechnology, a novel field concerned with manipulating materials at nanoscale with dimension of smaller than 100 nm, has recently gained attention. Nanoparticles have a high surface area to volume ratio, making them an important driver in various fields due to their unique properties at nanoscale. Similarly, fillers also exhibit high activity and chemical stability (Yang et al., 2010). Krawiec and colleagues discovered that the size of filler particles is a crucial factor that determines the conductivity of polymer electrolytes (Krawiec et al., 1995). They reported that adding nanosized Al_2O_3 to polymer electrolytes increased the conductivity by about an order of magnitude compared to that of micrometer-sized Al_2O_3. This is attributed to the small size of the fillers that enhance the uniformity and electrochemical properties of the sample. The higher conductivity of samples with the addition of nanoscale fillers is also due to the rapid development of the space charge region between the grains in the system (Mei et al., 2008).

Fillers can be broadly categorized as either inorganic or organic. Some examples of inorganic fillers are calcium carbonate, fly ash, mica, clay, manganese oxide (MnO_2), TiO_2, ZrO_2, SiO_2, and alumina (Al_2O_3), while organic fillers include graphite fiber, aromatic polyamide, and cellulosic rigid rods (whiskers) (Samir et al., 2005). The growing interest in the growth of fillers has led to the creation of a new type of filler, the organic–inorganic hybrid, which combines both organic and inorganic phases. Examples of these hybrids include poly(cyclotri-phosphazene-co-4,40-sulfonyldiphenol) (PZS) microspheres (Zhang et al., 2010). Inorganic fillers can be further classified as either active or passive. Active fillers are materials that participate in the ionic conduction process, such as lithium-nitrogen (Li_2N), lithium aluminate ($LiAlO_2$), and ceramics-containing lithium, titanium nitrides (TiN_x), titanium carbides (TiC_x), and titanium carbonitrides (TiC_xN_y) (Ishkov and Sagalakov; 2005). In contrast, passive fillers do not have a role in the transport mechanism of the charge carrier (Adebahr et al., 2003; Giffin et al., 2012).

Various types of inorganic ceramic fillers are utilized in the preparation of polymer electrolytes. These include inert metal oxides such as titanium dioxide (TiO_2), zirconium dioxide (ZrO_2), alumina (Al_2O_3), and manganese dioxide (MnO_2), as well as other substances like treated silica (SiO_2), molecular sieves and zeolites, rare earth oxides, ferroelectric materials, solid superacids, nanoclay such as montmorillonite (MMT), carbon nanotubes (CNTs), and heteropolyacids such as silicotungstic acid (SiWA) and phosphotungstic acid (Jung et al., 2009; Noto et al., 2012; Zapata et al., 2012). Fillers can enhance the mechanical strength of polymer electrolytes, promote ionic conduction, and improve electrochemical properties while providing enhanced membrane stability (Zhang et al., 2010; Lue et al., 2008; Polu and Kumar, 2013). In addition, fillers offer several other benefits in polymer electrolyte development such as improving interfacial stability between electrode and electrolyte, reducing the glass transition temperature (T_g) of polymer membrane, increasing cationic diffusivity, improving the physical and morphological properties of polymer matrix, lowering interfacial resistance, reducing water retention, enhancing thermal stability, and improving the long-term electrochemical stability of polymer electrolytes (Krawiec et al., 1995; Samir et al., 2005; Kim et al., 2002; Saikia et al., 2009; Hammami et al., 2013; Jian-Hua et al., 2008).

Adding nanofillers to amorphous polymers allows them to maintain their liquid-like characteristics, resulting in faster ionic mobility at the microscopic level. This behavior not only improves the performance of energy storage devices by reducing capacity fading but also increases the electrochemical stability and long-term cycle life, as reported in several studies using CPEs or NCPEs (Raghava et al., 2008; Yang et al., 2012). For example, PEG-magnesium acetate ($Mg(CH_3COO)_2$) containing 10 wt.% of alumina achieved a maximum ionic conductivity of 3.45×10^{-6} S/cm, leading to a fabricated battery with a current density of 13.91 $\mu A/cm^2$, discharge capacity of 1.721 mA h, power density of 13.14 mW/kg, and energy density of 1.84 W h/kg with an open circuit voltage (OCV) of 1.85 V (Jung et al., 2009). In another study, nanoscale TiO_2 was distributed into a porous membrane made up of two copolymers, PVdF-co-HFP and poly (ethylene oxide-co-ethylene carbonate) (P(EO-co-EC)). Ionic conductivity as high as 5.1×10^{-5} S/cm was achieved at room temperature, slightly higher than the polymer electrolyte without the addition of TiO_2 nanoparticles. Jeon et al. reported these findings in 2006 (Jeon et al., 2006).

Ketabi and Lian examined the impact of amorphous SiO_2 nanofillers on the ionic conductivity and crystallinity of PEO-based electrolytes (Ketabi and Lian, 2013). They found that the ionic conductivity of the polymer-ionic liquid electrolyte system, which consisted of 1-ethyl-3-methylimidazolium hydrogensulfate (EMIHSO$_4$) and nanosized SiO_2, was able to achieve 2.15 mS/cm. This represents a more than twofold increase over the electrolyte that did not contain fillers (Nordström et al., 2012). SiO_2 is a hydrophilic compound and prefers to form silanol (Si–OH) bonding on the surface of the silicon atom of silica as illustrated in Figure 5.8 (Ahmad et al., 2005).

Formation of hydrogen bonding between SiO_2 grains is favorable with the presence of Si–OH bond when a sufficient amount of SiO_2 is embedded into the polymer matrix. This will lead to the formation of a three-dimensional network whereby silica is immobilized in the liquid electrolyte (Ahmad et al., 2005). The charge carriers are thus accumulated on the silica surface as shown in Figure 5.9a. Therefore, the charge carriers are expelled from the surface easily when the hydrogen bonding between SiO_2 grains is formed as illustrated in Figure 5.9b (Nordström et al., 2012). As a result, the charge carriers can transport freely in the electrolyte.

Johan and Ting incorporated a novel nano-filler, specifically nanosized MnO_2, into a PEO-LiCF$_3$SO$_3$-DBP polymer system. The resulting NCPE achieved an optimal

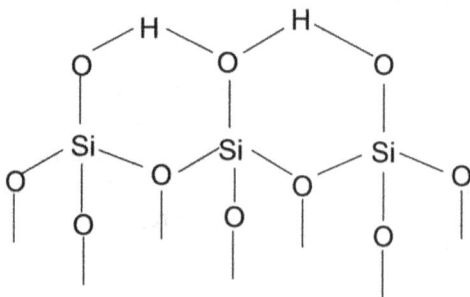

FIGURE 5.8   The formation of three-dimensional network with the silanol bonding.

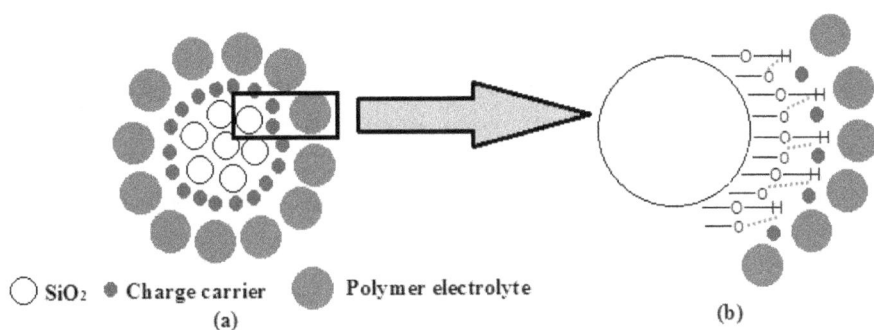

○ SiO₂ ● Charge carrier ⬤ Polymer electrolyte

(a)   (b)

FIGURE 5.9 (a) The immobilization of silica in the electrolyte and (b) the formation of hydrogen bonding between silica grains. Adapted from Nordström et al. (2012).

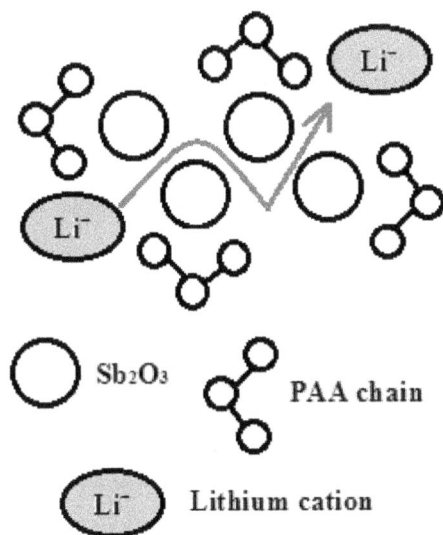

○ Sb₂O₃   PAA chain

Li⁻ Lithium cation

FIGURE 5.10 Formation of a conducting pathway between Sb₂O₃ particles. Adapted from Kam et al. (2014).

conductivity of 4.2×10^{-4} S/cm at room temperature when the MnO_2 was dispersed at a concentration of 12 wt.% (Johan and Ting, 2011). A new material namely antimony trioxide (Sb_2O_3) is also investigated by our peer (Kam et al., 2014). We found out that the ionic conductivity of poly(acrylic acid) (PAA)–based polymer electrolytes is increased rapidly at high concentration of fillers as reported in our publication. The filler grains could be distributed evenly throughout the film when more fillers are added to the polymer system. This presents an extra conducting path (also known as tunnel) which is accessible for ionic transportation around the surface of particles. As a result, the mobile charge carriers could diffuse within the vicinity of grains through this conducting trail as illustrated in Figure 5.10. The ionic conductivity of some selected CPEs is listed in Table 5.2.

TABLE 5.2

The Ionic Conductivity of Selected Composite Polymer Electrolytes

Polymer Electrolyte System	Ionic Conductivity (S/cm)	References
PEO-LiBF$_4$-MMT	1.60×10^{-6} S/cm at 303 K	Krawiec et al. (1995)
PEO-LiN(CF$_3$SO$_3$)$_2$-SiO$_2$	1.40×10^{-4} S/cm at ambient temperature	Capiglia et al. (1999)
(PEO)$_9$ LiN(CF$_3$SO$_3$)$_2$-Al$_2$O$_3$	2.20×10^{-4} S/cm at 298 K	Jayathilaka et al. (2002)
(PEO)$_9$Cu(CF$_3$SO$_3$)$_2$-Al$_2$O$_3$	9.07×10^{-7} S/cm at 301 K	Dissanayake et al. (2005)
PVdF-co-HFP-lithium fluoroalkyl phosphate (LiFAP)-(EC+DEC)-SiO$_2$	1.13×10^{-3} S/cm at ambient temperature	Aravindan and Vickraman (2007)
(PEO)$_9$LiCF$_3$SO$_3$-EC-Al$_2$O$_3$	1.50×10^{-4} S/cm at ambient temperature	Pitawala et al. (2007)
PVA-KOH-TiO$_2$	0.048 S/cm at 303 K	Yang et al. (2008)
PEO-LiClO$_4$-hydrotalcite	1.10×10^{-5} S/cm at ambient temperature	Borgohain et al. (2010)
PEO-LiClO$_4$-PZS	1.20×10^{-5} S/cm at ambient temperature	Zhang et al. (2010)
PMMA-LiCF$_3$SO$_3$-EC-Al$_2$O$_3$	2.05×10^{-4} S/cm at ambient temperature	Chew and Tan (2011)
PMMA-LiCF$_3$SO$_3$-EC-SiO$_2$	2.15×10^{-5} S/cm at ambient temperature	Chew and Tan (2011)
PVA-KOH-SiO$_2$	0.035 S/cm at 303 K	Yang et al. (2011)
Poly(vinylphosphonic acid) (PVPA)-sulfated TiO$_2$	0.03 S/cm at 423 K	Aslan and Bozkurt (2012)
Nafion-sulfated ZrO$_2$	3.00×10^{-3} S/cm at 393 K	Griffin et al. (2012)
PEO-LiPF$_6$-EC-amorphous CNTs	1.30×10^{-3} S/cm at ambient temperature	Ibrahim and Johan (2012)
Poly(ethyleneimine) (PEI)-LiN(CF$_3$SO$_3$)$_2$-SiO$_2$	3.80×10^{-5} S/cm at ambient temperature	Pehlivan et al. (2012)
Polysaccharide (agarose)-LiI-I$_2$-TiO$_2$	5.12×10^{-4} S/cm at ambient temperature	Raghava et al. (2008)
PEO-LiClO$_4$-Al$_2$O$_3$	6.45×10^{-4} S/cm at 293 K	Masoud et al. (2013)
PVdF-LiPF$_6$-(EC+DEC)-MMT	3.08×10^{-3} S/cm at ambient temperature	Prasanth et al. (2013)
Poly(ethylene oxide-co-2-(2-methoxyethoxy)ethyl glycidylether) P(EO/EM$_2$)-LiClO$_4$-TiO$_2$	1.40×10^{-5} S/cm at 303 K	Tominaga and Endo (2013)
PEO-tetraethylammonium tetrafluoroborate (TEABF$_4$) salt-organically modified nanoclay (MNclay)	1.0×10^{-4} S/cm at 303 K	Sivaraman et al. (2015)
PVA-PMMA-NH$_4$SCN-nanosized multiferrioc (BiFeO$_3$) (BFO)	2.9×10^{-2} S/cm at 293 K	Nileshrai et al. (2015)

Although the addition of fillers is a valuable approach for enhancing ionic conduction, the degree of improvement is currently limited. Based on the findings, the combined effect of plasticizer and filler can have a good impact on the ionic conductivity of polymer electrolytes.

5.4   FUTURE ASPECTS

5.4.1   Liquid Crystal Polymer Electrolytes (LCPEs)

Liquid crystal polymer electrolytes (LCPEs) represent a novel type of polymer electrolytes that replace common polymers with liquid crystal polymers (LCPs) as host polymers (Álvarez Moisés et al., 2023). In conventional polymer electrolytes, the transport of ions is achieved through segmental motions of the host polymers, leading to a significant increase in ionic conductivity above the glass transition temperature (T_g) only (Imrie et al., 2004). However, lowering the T_g to sub-ambient temperatures greatly reduces the mechanical stability of these electrolytes (Imrie and Ingram, 2000). To address this issue, LCPs have been developed, which exhibit high ionic conductivity in both glassy and liquid crystal phases (Imrie et al., 2004). Unlike amorphous polymer complexes, where ionic motion is coupled to structural relaxations, LCPs' liquid crystalline behavior can decouple the ion mobility from the structural relaxations (McHattie et al., 1998). Furthermore, LCPs offer additional benefits due to a combination of anisotropic and exceptional bulk properties, along with new possibilities for polymer processing (Park et al., 2010).

LCPs are polymers that contain mesogen, which are the fundamental functional groups of liquid crystals that promote structural order in crystals. These LCPs can be classified into two main categories based on whether the mesogens are located in the backbone or side chain of the polymer. These categories include main chain-liquid crystal polymers (MCLCPs) and side chain-liquid crystal polymers (SCLCPs) (refer Figure 5.11).

LCPs that contain mesogenic units in the backbone are referred to as MCLCPs. The mesomorphic properties of MCLCPs depend on the polymer's chain flexibility and structure. By incorporating mesogenic units into an ionically conducting polymer backbone, MCLCP electrolytes can be created (McHattie et al., 1998).

In contrast, SCLCPs are LCPs that have mesogenic groups in the side chain. The preparation of SCLCPs requires three main components: the polymer backbone, mesogenic units, and a space connecting the mesogenic units (Felipe, 2009).

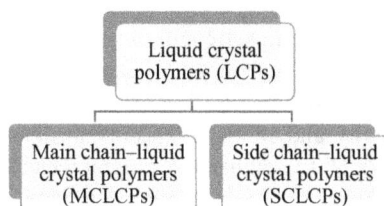

FIGURE 5.11   Two main types of liquid crystal polymers.

When an ion-coordinating cyclic macromolecule is inserted into the mesogenic units that are attached to the polymer backbone through flexible spacers, SCLCP electrolytes can be produced (McHattie et al., 1998). The mesomorphic properties of SCLCPs are influenced by various structural factors such as the flexibility of the polymer backbone, the repeating unit length, and the spacer length between the side chains of the polymer. In SCLCPs, the molecular mobility of the polymeric backbone and mesogenic groups are linked through the flexible spacer, which requires the decoupling of ionic motion from structural relaxation for effective ionic transport in polymer electrolytes. Increasing the mentioned structural factors can enhance the degree of decoupling, ultimately leading to a decrease in the T_g of polymer electrolytes and promoting the ionic hopping mechanism within the polymer matrix.

The development of ionically conducting LCPEs has been significantly advanced by Imrie and her research group (Imrie and Ingram, 2000; Imrie et al., 2004; McHattie et al., 1998). McHattie et al. created a new mesogenic liquid crystalline side-chain polymer with predominantly PEO backbone and mesogenic groups attached as pendants through flexible alkyl spacers (McHattie et al., 1998). PEO-LiClO$_4$-based liquid crystalline systems exhibit higher ionic conductivity and different conductivity behavior compared to pristine PEO-LiClO$_4$ polymer complexes. Tong et al. synthesized novel star-branched amphiphilic liquid crystal copolymers based on PEO with cyanobiphenyl mesogenic pendants (MA$_x$LC) using atom transfer radical polymerization (ATRP) (Tong et al., 2012). The insertion of mesogenic groups significantly increased the ionic conductivity of polymer electrolytes by providing an efficient ion-conducting pathway and suppressing polymer crystallization to promote movement (Tong et al., 2012). LCPEs have also been employed in electrochemical devices, with Park et al. fabricating dye-sensitized solar cells (DSSCs) using a new series of SCLCPs that achieved a maximum power conversion efficiency (PCE) of 4.11% and better photovoltaic performance (Park et al., 2010). Extensive research has been conducted to understand the impact of the liquid crystalline region on the ionic conductivity of polymer electrolytes for use in electrochemical devices (Chai et al., 2021).

5.4.2 Liquid Crystals-Embedded Polymer Electrolytes

Polymer electrolytes containing liquid crystals (LCs) are a new type of SPE that offers promising characteristics for use in electrochemical devices. LCs, which can form a mesophase with desirable properties such as high charge carrier mobility and conductivity, are a potential replacement for volatile liquids in electrolytes. The alignment of LCs in polymer electrolytes can improve the photovoltaic performance of DSSCs by increasing the ordering strength of the polymeric material, which provides a better pathway for charge carriers. LCs can also initiate the iodide exchange reaction and increase the chain mobility in the polymer electrolytes, resulting in improved ionic conductivity and short-circuit current density (J_{sc}). Several studies have reported on the use of LC-embedded polymer electrolytes in DSSCs, with increased PCE observed when the LC is aligned compared to when it is not aligned or in conventional DSSCs. Further research is needed to fully understand the effects of the LC region on the ionic conductivity of these polymer electrolytes.

There are literature works discussing polymer electrolyte system that incorporates LC and its application in DSSCs (Ahn et al., 2012; Kim et al., 2010; Vijayakumar et al., 2009). The study shows that the addition of E7 LC to PVdF-co-HFP-based polymer electrolytes significantly enhances the ionic conductivity of the GPE, with a value of 2.9 mS/cm at ambient temperature, 37% higher than the polymer gel electrolyte without E7. The assembled DSSCs using this LC-embedded polymer electrolyte show comparable performance to those using liquid electrolytes, with a PCE of 6.82% at 1 sun intensity. Karim et al. also fabricated DSSCs using LC-embedded polymer electrolytes, achieving a PCE of 4.70% at 1 sun intensity, which is much higher than the PAN-based polymer electrolyte (Karim et al., 2010). Although the effect of LCs on the ionic conductivity of polymer electrolytes is not fully understood, the incorporation of LC in polymer electrolytes is a promising approach to improve their ionic conductivity in the future.

5.5   CONCLUSIONS

Polymer electrolytes are a viable alternative to hazardous liquid electrolytes. The requirements of an ideal polymer electrolyte are high ionic conductivity (> mS/cm), excellent thermal stability, wide electrochemical potential window, and superior electrochemical performances in cell fabrication. Development and improvement are still needed to overcome the current shortcomings of SPEs. Perhaps, SPE can be a good replacement of the current liquid electrolytes. Fuel cells offer a potentially non-polluting and renewable way to generate electricity. PEMFCs offer several advantages over the batteries and supercapacitors. Fabrication of a fuel cell with high power density, long durability, and cost-effective properties is essential for the economic application of sustainable energy technologies. The research and development (R&D) of PEMFCs is still essential to harness the capabilities of hydrogen for energy production to the maximum threshold and operational advantages prior to commercialization. Innumerable challenges and technological breakthroughs are still the obstacles in the R&D of PEMFCs.

ACKNOWLEDGMENT

This work was partly supported by the Fundamental Research Grant Scheme (FRGS/1/2019/STG07/TARUC/02/1) from the Ministry of Higher Education of Malaysia.

REFERENCES

Adebahr J, Byrne N, Forsyth M, MacFarlane DR, Jacobsson P (2003) Enhancement of ion dynamics in PMMA-based gels with addition of TiO_2 nano-particles. *Electrochim Acta* 48: 2099–2103.
Ahmad A, Isa KB, Osman Z (2011) Conductivity and structural studies of plasticized polyacrylonitrile (PAN)-lithium trifate polymer electrolyte films. *Sains Malaysiana* 40: 691–694.
Ahmad S, Ahmad S, Agnihotry SA (2005) Nanocomposite electrolytes with fumed silica in poly(methyl methacrylate): Thermal, rheological and conductivity studies. *J Power Sources* 140: 151–156.

Ahmad S, Deepa M, Agnihotry SA (2008) Effect of salts on the fumed silica-based composite polymer electrolytes. *Solar Energy Mater Solar Cells* 92: 184–189.

Ahn SK, Ban T, Sakthivel P, Lee JW, Gal Y-S, Lee J-K, Kim M-R, Jin S-H (2012) Development of dye-sensitized solar cells composed of liquid crystal embedded, electrospun poly(vinylidene fluoride-co-hexafluoropropylene) nanofibers as polymer gel electrolytes. *ACS Appl Mater Interfaces* 4: 2096–2100.

Akiyama Y, Sodaye H, Shibahara Y, Honda Y, Tagawa S, Nishijima S (2010) Study on gamma-ray-induced degradation of polymer electrolyte by pH titration and solution analysis. *Polym Degrad Stab* 95: 1–5.

Ali AMM, Yahya MZA, Bahron H, Subban RHY, Harun MK, Atan I (2007) Impedance studies on plasticized PMMA-LiX (X: CF_3SO_3-, $N(CF_3SO_2)^{2-}$) polymer electrolytes. *Mater Lett* 61: 2026–2029.

Álvarez Moisés I, Innocenti A, Somville M, Notredame B, Passerini S, Gohy, JF (2023). Liquid crystals as additives in solid polymer electrolytes for lithium metal batteries. *MRS Adv* 8: 797.

Aravindan V, Vickraman P (2007) Polyvinylidenefluoride-hexafluoropropylene based nano-composite polymer electrolytes (NCPE) complexed with $LiPF_3(CF_3CF_2)_3$. *Eur Polym J* 43: 5121–5127.

Armand MB (1986) Polymer electrolytes. *Ann Rev Mater Res* 16: 245–261.

Arof AK, Ramesh S (2000) Electrical conductivity studies of poly(vinylchloride) based electrolytes with double salt system. *Solid State Ionics* 136–137: 1197–1200.

Aruchamy K, Ramasundaram S, Divya S, Chandran M, Yun K, OhTH (2023). Gel polymer electrolytes: Advancing solid-state batteries for high-performance applications. *Gels* 9: 585.

Aslan A, Bozkurt A (2012) Nanocomposite polymer electrolyte membranes based on poly(vinylphosphonic acid)/sulfated nano-titania. *J Power Sources* 217: 158–163.

Baskaran R, Selvasekarapandian S, Kuwata N, Kawamura J, Hattori T (2006a) Conductivity and thermal studies of blend polymer electrolytes based on PVAc-PMMA. *Solid State Ionics* 177: 2679–2682.

Baskaran R, Selvasekarapandian S, Kuwata N, Kawamura J, Hattori T (2006b) Ac impedance, DSC and FT-IR investigations on (x)PVAc-(1- x)PVdF blends with $LiClO_4$. *Mater Chem Phys* 98: 55–61.

Baskaran R, Selvasekarapandian S, Kuwata N, Kawamura J, Hattori T (2007) Structure, thermal and transport properties of PVAc-$LiClO_4$ solid polymer electrolytes. *J Physics Chem Solids* 68: 407–412.

Bella RSD, Hirankumar G, Devaraj P (2014) Characterization of plasticized proton conducting polymer electrolyte and its application in primary proton battery. *Int J ChemTech Res* 6: 5372–5377.

Borgohain MM, Joykumar T, Bhat SV (2010) Studies on a nanocomposite solid polymer electrolyte with hydrotalcite as a filler. *Solid State Ionics* 181: 964–970.

Bouridah A, Dalard F, Deroo D, Cheradame H, LeNest JF (1985) Poly(dimethylsiloxane)-poly(ethylene oxide) based polyurethane networks used as electrolytes in lithium electrochemical solid state batteries. *Solid State Ionics* 15: 233–240.

Braun D, Cherdron H, Rehahn M, Ritter H, Voit B (2005) Methods and techniques for synthesis, characterization, processing, and modification of polymers. In: *Polymer Synthesis: Theory and Practice*, Springer, Berlin, Germany, Vol. 1, pp 33–147.

Bruce PG, Vincent CA (1993) Polymer electrolytes. *J Chem Soc Faraday Trans* 89: 3187–3203.

Capiglia C, Mustarelli P, Quartarone E, Tomasi C, Magistris A (1999) Effects of nanoscale SiO_2 on the thermal and transport properties of solvent-free, poly(ethylene oxide) (PEO)-based polymer electrolytes. *Solid State Ionics* 118: 73–79.

Chai MN, Isa MIN (2014) Electrical study of plasticized carboxy methylcellulose based solid polymer electrolyte. *Int J Phys Sci* 9: 397–401.

Chai S, Xu F, Zhang R, Wang X, Zhai L, Li X, Qian HJ, Wu L, Li H (2021). Hybrid liquid-crystalline electrolytes with high-temperature-stable channels for anhydrous proton conduction. *J Am Chem Soc* 143, 21433–21442.

Cheng H, Zhu C, Huang B, Lu M, Yang Y (2007) Synthesis and electrochemical characterization of PEO-based polymer electrolytes with room temperature ionic liquids. *Electrochim Acta* 52: 5789–5794.

Chew KW, Tan KW (2011) The effects of ceramic fillers on PMMA-based polymer electrolyte salted with lithium triflate, $LiCF_3SO_3$. *Int J Electrochem Sci* 6: 5792–5801.

Choi NS, Lee YG, Park JK, Ko JM (2001) Preparation and Electrochemical characteristics of the plasticized polymer electrolyte based on P(VdF-co-HFP)/PVAc blend. *Electrochim Acta* 46: 1581–1586.

Choi NS, Park JK (2001) New polymer electrolytes based on PVC/PMMA blend for plastic lithium-ion batteries. *Electrochim Acta* 46 1453–1459.

Cowie JMG, Spence GH (1999) Novel single ion, comb-branched polymer electrolytes. *Solid State Ionics* 123: 233–242.

Cui ZY, Xu YY, Zhu LP, Wei XZ, Zhang CF, Zhu BK (2008) Preparation of PVDF/PMMA blend microporous membranes for lithium ion batteries via thermally induced phase separation process. *Mater Lett* 62: 3809–3811.

Damle R, Kulkarni PN, Bhat SV (2008) Study of effect of composition, irradiation and quenching on ionic conductivity in $(PEG)_x:NH_4NO_3$ solid polymer electrolyte. *Bull Mater Sci* 31: 869–876.

Deshpande VK (2009) Science and technology of glassy solid electrolytes. *IOP Conf Ser Mater Sci Eng* 2: 12011–12015.

Ding J, Chuy C, Holdcroft S (2002) Enhanced conductivity in morphologically controlled proton exchange membranes: Synthesis of macromonomers by SFRP and their incorporation into graft polymers. *Macromolecules* 35: 1348–1355.

Dissanayake MAKL, Jayathilaka PARD, Bokalawela RSP (2005) Ionic conductivity of PEO_9: $Cu(CF_3SO_3)_2:Al_2O_3$ nano-composite solid polymer electrolyte. *Electrochim Acta* 50: 5602–5605.

Eliasson H, Albinsson I, Mellander BE (2000) Conductivity and dielectric properties of $AgCF_3SO_3$-PPG. *Mater Res Bull* 35: 1053–1065.

Famprikis T, Canepa P, Dawson JA, Islam MS and Masquelier C (2019). Fundamentals of inorganic solid-state electrolytes for batteries. *Nat Mater* 18, 1278–1291.

Fan L, Dang Z, Nan C-W (2002) Thermal, electrical and mechanical properties of plasticized polymer electrolytes based on PEO/P(VDF-HFP) blends. *Electrochim Acta* 48: 205–209.

Fan L, Nan C- W, Zhao S (2003) Effect of modified SiO_2 on the properties of PEO-based polymer electrolytes. *Solid State Ionics* 164: 81–86.

Felipe AM (2009) Preparation and characterisation of new materials for electrolytes used in direct methanol fuel cells. Doctoral Dissertation, University of Aberdeen and Univesitat Politècnica de València, Valencia, Spain. https://riunet.upv.es/bitstream/handle/10251/8327/tesisUPV3183.pdf?

Fenton DE, Parker JM, Wright PV (1973) Complexes of alkali metal ions with poly(ethylene oxide). *Polym* 14: 589.

Fish D, Khan IM, Smid J (1988) Conductivity of solid complexes of lithium perchlorate with poly {(ω-methoxyhexa(oxyethylene)ethoxy)methylsiloxane}. *Makromolekul Chem Rapid Comm* 7: 115–120.

Fusco FA, Tuller HL (1989) Fast ion transport in glasses In Laskar AL, Chandra S (eds) *Superionic Solids and Solid Electrolytes Recent Trends*. Academic Press, San Diego, pp. 43–110.

Ganesan S, Muthuraaman B, Mathew V, Madhavan J, Maruthamuthu P, Suthanthiraraj SA (2008) Performance of a new polymer electrolyte incorporated with diphenylamine in nanocrystalline dye-sensitized solar cell. *Sol Energy Mater Sol C* 92: 1718–1722.

Ghosal S, Ray R, Ballabh TK, Tarafdar S (2013) Study of diffusion and conduction in gamma irradiated solid polymer electrolytes by fractal model structure. *Indian J Pure Appl Phys* 51: 324–327.

Giffin GA, Piga M, Lavina S, Navarra MA, D'Epifanio A, Scrosati B, Noto VD (2012) Characterization of sulfated-zirconia/Nafion composite membranes for proton exchange membrane fuel cells. *J Power Sources* 198: 66–75.

Gray FM (1991) *Solid Polymer Electrolytes: Fundamentals of Technological Applications.* Wiley-VCH, New York.

Gray FM (1997) *Polymer Electrolytes.* The Royal Society of Chemistry, London.

Guilherme LA, Borges RS, Moraes EMS, Silva GG, Pimenta MA, Marletta A, Silva RA (2007) Ionic conductivity in polyethylene-b-poly(ethylene oxide)/lithium perchlorate solid polymer electrolytes. *Electrochim Acta* 53: 1503–1511.

Hammami R, Ahamed Z, Charradi K, Beji Z, Assaker IB, Naceur JB, Auvity B, Squadrito G, Chtourou R (2013) Elaboration and characterization of hybrid polymer electrolytes Nafion-TiO$_2$ for PEMFCs. *Int J Hydrogen Energy* 38: 11583–11590.

Han HS, Kang HR, Kim SW, Kim HT (2002) Phase separated polymer electrolyte based on poly(vinyl chloride)/poly(ethyl methacrylate) blend. *J Power Sources* 112: 461–468.

Hayashi A, Noi K, Sakuda A, Tatsumisago M (2012) Superionic glass-ceramic electrolytes for room-temperature rechargeable sodium batteries. *Nat Comm* 3: 856–860.

Ibrahim S, Johan MR (2012) Thermolysis and conductivity studies of poly(Ethylene Oxide) (PEO) based polymer electrolytes doped with carbon nanotube. *Int J Electrochem Sci* 7: 2596–2615.

Ibrahim S, Yasin SMM, Ahmad R, Johan MR (2012) Conductivity, thermal and morphology studies of PEO based salted polymer electrolytes. *Solid State Sci* 14: 1111–1116.

Imrie CT, Ingram M (2000) Bridging the gap between polymer electrolytes and inorganic glasses: Side group liquid crystal polymer electrolytes. *Mol Cryst Liq Cryst* 347: 199–210.

Imrie CT, Inkster RT, Lu Z, Ingram M (2004) Discotic side group liquid crystal polymer electrolytes. *Mol Cryst Liq Cryst* 408: 33–43.

Ishkov AV, Sagalakov AM (2005) Effect of active fillers on properties of heat-resistant composites. *Russ J Appl Chem* 78: 1512–1516.

Jain N, Kumar A, Chauhan S, Chauhan SMS (2005) Chemical and biochemical transformations in ionic liquids. *Tetrahedron* 61: 1015–1060.

Jannasch P (2000) Synthesis of novel aggregating comb-shaped polyethers for use as polymer electrolytes. *Macromolecules* 33: 8604–8610.

Jayathilaka PARD, Dissanayake MAKL, Albinsson I, Mellander B-E (2002) Effect of nanoporous Al$_2$O$_3$ on thermal, dielectric and transport properties of the (PEO)$_9$LiTFSI polymer electrolyte system. *Electrochim Acta* 47: 3257–3268.

Jeon J-D, Kim M-J, Kwak S-Y (2006) Effect of addition of TiO$_2$ nanoparticles on mechanical properties and ionic conductivity of solvent-free polymer electrolytes based on porous P(VdF-HFP)/P(EO-EC) membranes. *J Power Sources* 162: 1304–1311.

Jian-Hua T, Gao P-f, Zhang Z-y, Luo W-h, Shan Z-q (2008) Preparation and performance evaluation of a Nafion-TiO$_2$ composite membrane for PEMFCs. *Int J Hydrogen Energy* 33: 5686–5690.

Johan MR, Ting LM (2011) Structural, thermal and electrical properties of nano manganese-composite polymer electrolytes. *Int J Electrochem Sci* 6: 4737–4748.

Jung S, Kim DW, Lee SD, Cheong M, Nguyen DQ, Cho BW, Kim HS (2009) Fillers for solid-state polymer electrolytes: Highlight. *Bull Korean Chem Soc* 30: 2355–2361.

Kam W, Liew C-W, Lim JY, Ramesh S (2014) Electrical, structural, and thermal studies of antimony trioxide-doped poly(acrylic acid)-based composite polymer electrolytes. *Ionics* 20: 665–674.

Kang X (2004) Nonaqueous liquid electrolytes for lithium-based rechargeable batteries. *Chem Rev* 104: 4303–4417.

Karan NK, Pradhan DK, Thomas R, Natesan B, Katiyar RS (2008) Solid polymer electrolytes based on polyethylene oxide and lithium trifluoromethanesulfonate (PEO-LiCF₃SO₃): Ionic conductivity and dielectric relaxation. *Solid State Ionics* 179: 689–696.

Karim MA, Song M, Park JS, Kim YH, Lee MJ, Lee JW, Lee CW, Cho Y-R, Gal Y-S, Lee JH, Jin S-H (2010) Development of liquid crystal embedded in polymer electrolytes composed of click polymers for dye-sensitized solar cell applications. *Dyes Pigm* 86: 259–265.

Ketabi S, Lian K (2013) Effect of SiO₂ on conductivity and structural properties of PEO-EMIHSO₄ polymer electrolyte and enabled solid electrochemical capacitors. *Electrochim Acta* 103: 174–178.

Kim H-K, Hur S-Y, Kang W-S, Lee G-D (2010) Homeotropic aligned liquid crystal molecules in dye-sensitized solar cells for high efficiency. *J Korean Phys Soc* 56: 1519–1522.

Kim KM, Park N-G, Ryu KS, Chang KS (2002) Characterization of poly(vinylidenefluoride-co-hexafluoropropylene)-based polymer electrolytes filled with TiO₂ nanoparticles. *Polym* 43: 3951–3957.

Kim KS, Park SY, Choi S, Lee H (2006) Ionic liquid-polymer gel electrolytes based on morpholinium salt and PVdF(HFP) copolymer. *J Power Sources* 155: 385–390.

Kitaura H, Hayashi A, Ohtomo T, Hama S, Tatsumisago M. (2011) Fabrication of electrode-electrolyte interfaces in all-solid-state rechargeable lithium batteries by using a supercooled liquid state of the glassy electrolytes. *J Mater Chem* 21: 118–124.

Kono M, Furuta K, Mori, S, Watanabe M, Ogata N (1993) Synthesis of polymer electrolytes based on poly(2-(2-methoxyethoxy)ethyl glycidyl ether) and their high ionic conductivity. *Polym Advan Technol* 4: 85–91.

Krawiec W, Jr Scanlon LG, Fellner JP, Vaia RA, Vasudevan S, Giannelis EP (1995) Polymer nanocomposites: A new strategy for synthesizing solid electrolytes for rechargeable lithium batteries. *J Power Sources* 54: 310–315.

Kudo T, Fueki K (1990) *Solid State Ionics*. VCH and Kodansha Ltd, Tokyo, Vol. 1, pp. 29–42.

Kuila T, Acharya H, Srivastava SK, Samantaray BK, Kureti S (2007) Enhancing the ionic conductivity of PEO based plasticized composite polymer electrolyte by LaMnO₃ nanofiller. *Mater Sci Eng B* 137: 217–224.

Kumar MS, Bhat DK (2009) Polyvinyl alcohol-polystyrene sulphonic acid blend electrolyte for supercapacitor application. *Phys B* 404: 1143–1147.

Kuo C-W, Li W-B, Chen P-R, Liao J-W, Tseng C-G, Wu T-Y (2013) Effect of plasticizer and lithium salt concentration in PMMA based composite polymer electrolytes. *Int J Electrochem Sci* 8: 5007–5021.

Li Z, Fu J, Zhou X, Gui S, Wei L, Yang H, Li H, Guo X (2023). Ionic conduction in polymer-based solid electrolytes. *Adv Sci* 10, 18p..

Liew C-W, Ong YS, Lim JY, Lim CS, Teoh KH, Ramesh S (2013) Effect of ionic liquid on semi-crystalline poly(vinylidene fluoride-co-hexafluoropropylene) solid copolymer electrolytes. *Int J Electrochem Sci* 8: 7779–7794.

Liew C-W, Ramesh S (2013) Studies on ionic liquid-based corn starch biopolymer electrolytes coupling with high ionic transport number. *Cellulose* 20: 3227–3237.

Liew C-W, Ramesh S (2014) Comparing triflate and hexafluorophosphate anions of ionic liquids in polymer electrolytes for supercapacitor applications. *Mater* 7: 4019–4033.

Lue SJ, Lee D-T, Chen J-Y, Chiu C-H, Hu C-C, Jean YC, Lai J-Y (2008) Diffusivity enhancement of water vapor in poly(vinyl alcohol)-fumed silica nano-composite membranes: Correlation with polymer crystallinity and free-volume properties. *J Membrane Sci* 325: 831–839.

Lyons LJ, Southworth BA, Stam D, Yuan C-H, West R (1996) Polymer electrolytes based on polysilane comb polymer. *Solid State Ionics* 91: 169–173.

Masoud EM, Bellihi A-A El, Bayoumy WA, Mousa MA (2013) Organic-inorganic composite polymer electrolyte based on PEO-LiClO$_4$ and nano-Al$_2$O$_3$ filler for lithium polymer batteries: Dielectric and transport properties. *J Alloy Compd* 575: 223–228.

McHattie GS, Imrie CT, Ingram MD (1998) Ionically conducting side chain liquid crystal polymer electrolytes. *Electrochim Acta* 43: 1151–1154.

Mei A, Wang X-L, Feng Y-C, Zhao S-J, Li G-J, Geng H-X, Lin Y-H, Nan C-W (2008) Enhanced ionic transport in lithium lanthanum titanium oxide solid state electrolyte by introducing silica. *Solid State Ionics* 179:, 2255–2259.

Michael MS, Jacob MME, Prabaharan SRS, Radhakrishana S (1997) Enhanced lithium ion transport in PEO-based solid polymer electrolytes employing a novel class of plasticizers. *Solid State Ionics* 98: 167–174.

Morales E, Acosta JL (1997) Thermal and electrical characterization of plasticized polymer electrolytes based on polyethers and polyphosphazene blends. *Solid State Ionics* 96: 99–106.

Nanda P, De SK, Manna S, De U, Tarafdar S (2010) Effect of gamma irradiation on a polymer electrolyte: Variation in crystallinity, viscosity and ion-conductivity with dose. *Nucl Instrum Methods Phys Res Sec B* 268: 73–78.

Nicotera I, Ranieri GA, Terenzi M, Chadwick AV, Webster MI (2002) A study of stability of plasticized PEO electrolytes. *Solid State Ionics* 146: 143–150.

Nileshrai, Markandaysingh, Abas M, Agarwal SL (2015) Experimental studies on blend based polymer nanocomposite electrolytes: PVA:PMMA:NH$_4$SCN:BFO System. *Int J Inno Appl Res* 3: 23–38.

Ning W, Xingxiang Z, Haihui L, Benqiao H (2009) 1-Allyl-3-methylimidazolium chloride plasticized-corn starch as solid biopolymer electrolytes. *Carbohydr Polym* 76: 482–484.

Nishimoto A, Agehara K, Furuya N, Watanabe T, Watanabe M (1999) High ionic conductivity of polyether-based network polymer electrolytes with hyperbranched side chains. *Macromolecules* 32: 1541–1548.

Nogueira VC, Longo C, Nogueira AF, Soto-Oviedo MA, Paoli M-AD (2006) Solid-state dye-sensitized solar cell: Improved performance and stability using a plasticized polymer electrolyte. *J Photochem Photobio A Chem* 181: 226–232.

Nordström J, Aguilera L, Matic A (2012) Effect of lithium salt on the stability of dispersions of fumed dilica in the ionic liquid BMImBF$_4$. *Langmuir* 28: 4080–4085.

Noto VD, Bettiol M, Bassetto F, Boaretto N, Negro E, Lavina S, Bertasi F (2012) Hybrid inorganic-organic nanocomposite polymer electrolytes based on Nafion and fluorinated TiO$_2$ for PEMFCs. *Int J Hydrogen Energy* 37: 6169–6181.

Noto VD, Lavina S, Giffin GA, Negro E, Scrosati B (2011) Polymer electrolytes: Present, past and future. *Electrochim Acta* 57: 4–13.

Osinska M, Walkowiak M, Zalewska A, Jesionowski T (2009) Study of the role of ceramic filler in composite gel electrolytes based on microporous polymer membranes. *J Membr Sci* 326: 582–588.

Pandey GP, Hashmi SA (2009) Experimental investigations of an ionic-liquid-based, magnesium ion conducting, polymer gel electrolyte. *J Power Sources* 187: 627–634.

Pandey GP, Hashmi SA (2013) Ionic liquid 1-ethyl-3-methylimidazolium tetracyanoborate-based gel polymer electrolyte for electrochemical capacitors. *J Mater Chem A* 1: 3372–3378.

Park JS, Kim YH, Song M, Kim C-H, Karim MA, Lee JW, Gal Y-S, Kumar P, Kang S-W, Jin S-H (2010) Synthesis and photovoltaic properties of side-chain liquid-crystal click polymers for dye-sensitized solar-cells application. *Macromol Chem Phys* 211: 2464–2473.

Park MJ, Choi I, Hong J, Kim O (2013) Polymer electrolytes integrated with ionic liquids for future electrochemical devices. *J Appl Polym Sci* 129: 2363–2376.

Patel M, Gnanavel M, Bhattacharyya AJ (2011) Utilizing an ionic liquid for synthesizing a soft matter polymer "gel" electrolyte for high rate capability lithium-ion batteries. *J Mater Chem* 21: 17419–17424.

Pehlivan İB, Granqvist CG, Marsal R, Georén P, Niklasson GA (2012) (PEI-SiO₂):(LiTFSI) nanocomposite polymer electrolytes:Ion conduction and optical properties. *Sol Energy Mater Sol C* 98: 465–471.

Pitawala HMJC, Dissanayake MAKL, Seneviratne VA (2007) Combined effect of Al₂O₃ nano-fillers and EC plasticizer on ionic conductivity enhancement in the solid polymer electrolyte (PEO)₉LiTf. *Solid State Ionics* 178: 885–888.

Polu AR, Kumar R (2013) Effect of Al₂O₃ ceramic filler on PEG-based composite polymer electrolytes for magnesium batteries. *Advan Mater Lett* 4: 543–547.

Pradhan DK, Samantaray BK, Choudhary RNP, Thakur AK (2005) Effect of plasticizer on structure-property relationship in composite polymer electrolytes. *J Power Sources* 139: 384–393.

Prasanth R, Shubha N, Hng HH, Srinivasan M (2013) Effect of nano-clay on ionic conductivity and electrochemical properties of poly(vinylidene fluoride) based nanocomposite porous polymer membranes and their application as polymer electrolyte in lithium ion batteries. *Eur Polym J* 49: 307–318.

Quartarone E, Mustarelli P (2011) Electrolytes for solid-state lithium rechargeable batteries: Recent advances and perspectives. *Chem Soc Rev* 40: 2525–2540.

Quartarone E, Mustarelli P, Magistris A (1998) PEO-based composite polymer electrolytes. *Solid State Ionics* 110: 1–14.

Ragavendran K, Kalyani P, Veluchamy A, Banumathi S, Thirunakaran R, Benedict TJ (2004) Characterization of plasticized PEO based solid polymer electrolyte by XRD and AC impedance methods. *Portugaliae Electrochim Acta* 22: 149–159.

Raghava P, Zhao X, Kim J-K, Manuel J, Chauhan GS, Ahn J-H, Nah C (2008) Ionic conductivity and electrochemical properties of nanocomposite polymer electrolytes based on electrospun poly(vinylidenefluoride-co-hexafluoropropylene) with nano-sized ceramic fillers. *Electrochim Acta* 54: 228–234.

Raghavan P, Zhao X, Manuel J, Chauhan GS, Ahn JH, Ryub HS, Ahn HJ, Kim KW, Nah C (2010) Electrochemical performance of electrospun poly(vinylidene fluoride-co-hexafluoropropylene)-based nanocomposite polymer electrolytes incorporating ceramic fillers and room temperature ionic liquid. *Electrochim Acta* 55: 1347–1354.

Raghavan V (2004) *Materials Science and Engineering: A First Course*. Prentice Hall of India Private Limited, New Delhi, Vol. 5, pp 332–354.

Rahaman MHA, Khandaker MU, Khan ZR, Kufian MZ, Noor ISM, Arof AK (2014) Effect of gamma irradiation on poly(vinyledene difluoride)-lithium bis(oxalato)borate electrolyte. *Phys Chem Chem Phys* 16: 11527–11537.

Rajendran S, Mahendran O, Kannan R (2002) Characterisation of ((1-) PMMA-PVdF) polymer blend electrolyte with Li⁺ ion. *Fuel* 81: 1077–1081.

Rajendran S, Prabhu MR, Rani MU (2008) Ionic conduction in poly(vinyl chloride)/poly(ethyl methacrylate)-based polymer blend electrolytes complexed with different lithium salts. *J Power Sources* 180: 880–883.

Rajendran S, Sivakumar M, Subadevi R (2004) Investigations on the effect of the various plasticizers in PVA-PMMA solid polymer blend electrolytes. *Mater Lett* 58: 641–649.

Rajendran S, Sivakumar P (2008) An investigation of PVdF/PVC-based blend electrolytes with EC/PC as plasticizers in lithium battery applications. *Phys B* 403: 509–516.

Rajendran S, Uma T, Mahalingam T (2000) Conductivity studies on PVC±PMMA±LiAsF6±DBP polymer blend electrolyte. *Eur Polym J* 36: 2617–2620.

Ramesh S, Arof AK (2001) Ionic conductivity studies of plasticized poly(vinyl chloride) polymer electrolytes. *Mater Sci Eng B* 85: 11–15.

Ramesh S, Chao LZ (2011) Investigation of dibutyl phthalate as plasticizer on poly(methyl methacrylate)-lithium tetraborate based polymer electrolytes. *Ionics* 17: 29–34.

Ramesh S, Liew C-W (2013) Development and investigation on PMMA-PVC blend-based solid polymer electrolytes with LiTFSI as dopant salt. *Polym Bull* 70: 1277–1288.

Ramesh S, Liew C-W, Arof AK (2011) Ion conducting corn starch biopolymer electrolytes doped with ionic liquid 1-butyl-3-methylimidazolium hexafluorophosphate. *J Non-Cryst Solids* 357: 3654–3660.

Ramesh S, Liew C-W, Lau PY, Ezra M (2012) Characterization of high molecular weight poly (vinyl chloride)-lithium tetraborate electrolyte plasticized by propylene carbonate. In: Luqman M (ed) *Recent Advances in Plasticizers*. In Tech, Crotia, Rijeka, pp. 165–190.

Ramesh, S, Liew C-W, Morris E, Durairaj R (2010) Effect of PVC on ionic conductivity, crystallographic structural, morphological and thermal characterizations in PMMA-PVC blend-based polymer electrolytes. *Thermochim Acta* 511: 140–146.

Ramesh S, Liew C-W, Ramesh K (2011) Evaluation and investigation on the effect of ionic liquid onto PMMA-PVC gel polymer blend electrolytes. *J Non-Cryst Solids* 357: 2132–2138.

Ramesh S, Liew C-W, Ramesh K (2013) Ionic conductivity: Dielectric behavior and HATR-FTIR analysis onto PMMA-PVC binary solid polymer blend electrolytes. *J Appl Polym Sci* 127: 2380–2388.

Reiter J, Vondrak J, Michalek J, Micka Z (2006) Ternary polymer electrolytes with 1-methylimidazole based ionic liquids and aprotic solvents. *Electrochim Acta* 52: 1398–1408.

Sagaidak I, Huertas G, Van NhienAN, Sauvage F (2016). Towards renewable iodide sources for electrolytes in dye-sensitized solar cells. *Energies* 9, 241.

Saikia D, Chen-Yang YW, Chen YT, Li YK, Lin SI (2009) ^{7}Li NMR spectroscopy and ion conduction mechanism of composite gel polymer electrolyte: A comparative study with variation of salt and plasticizer with filler. *Electrochim Acta* 54: 1218–1227.

Samir MASA, Alloin F, Sanchez J-Y, Dufresne A (2005) Nanocomposite polymer electrolytes based on poly(oxyethylene) and cellulose whiskers. *Polímeros: Ciência e Tecnologia* 15: 109–113.

Sekhar PC, Kumar PN, Sharma AK (2012) Effect of plasticizer on conductivity and cell parameters of (PMMA+NaClO$_4$) polymer electrolyte system. *IOSR J Appl Phys* 2: 1–6.

Sekhon SS, Krishnan P, Singh B, Yamada K, Kim CS (2006) Proton conducting membrane containing room temperature ionic liquid. *Electrochim Acta* 52: 1639–1644.

Sharma R, Sil A, Ray S (2012) Characterization of Plasticized PMMA-LiClO$_4$ Solid Polymer Electrolytes. *Advan Mater Res* 585: 185–189.

Singh PK, Kim K-W, Rhee H-W (2009) Development and characterization of ionic liquid doped solid polymer electrolyte membranes for better efficiency. *Synth Met* 159: 1538–1541.

Sinha M, Goswami MM, Mal D, Middya TR, Tarafdar S, De U, Chaudhuri SK, Das D (2008) Effect of gamma irradiation on the polymer electrolyte PEO-NH$_4$ClO$_4$. *Ionics* 14: 323–327.

Sirisopanaporn C, Fernicola A, Scrosati B (2009) New ionic liquid-based membranes for lithium battery application. *J Power Sources* 186: 490–495.

Sivakumar M, Subadevi R, Rajendran S, Wu H-C, Wu N-L (2007) Compositional effect of PVdF-PEMA blend gel polymer electrolytes for lithium polymer batteries. *Eur Polym J* 43: 4466–4473.

Sivaraman P, Shashidhara K, Thakur AP, Samui AB, Bhattacharyya AR (2015) Nanocomposite solid polymer electrolytes based on polyethylene oxide, modified nanoclay, and tetraethylammonium tetrafluoroborate for application in solid-state supercapacitor. *Polym Eng Sci* 55: 1536–1545.

Smart L, Moore EA (2005) Defects and nonstoichiometry. In: *Solid State Chemistry: An Introduction*. Taylor & Francis, New York, Vol. 4, pp 201–270.

Song Y, Wu S, Jing X, Sun J, Chen D (1997) Thermal, mechanical and ionic conductive behaviour of gamma-radiation induced PEO/PVDF(SIN)-LiClO$_4$ polymer electrolyte system. *Radiat Phys Chem* 49: 541–546.

Soo PP, Huang B, Jang Y, Chiang Y-M, Sadoway DR, Mayes AM (1999) Rubbery block copolymer electrolytes for solid-state rechargeable lithium batteries. *J Electrochem Soc*, 146: 32–37.

Souquet J-L, Nascimento MLF, Rodrigues ACM (2010) Charge carrier concentration and mobility in alkali silicates. *J Chem Phys* 132, 034704-1–034704-7.

Stephan AM, Nahm KS (2006) Review on composite polymer electrolytes for lithium batteries. *Polym* 47: 5952–5964.

Stephan AM, Saito Y, Muniyandi N, Renganathan NG, Kalyanasundaram S, Elizabeth RN (2002) Preparation and characterization of PVC/PMMA blend polymer electrolytes complexed with LiN(CF$_3$SO$_2$)$_2$. *Solid State Ionics* 148: 467–473.

Steven MP (1999) *Polymer Chemistry: An Introduction*. Oxford University Press, New York.

Sukeshini AM, Kulkarni AR, Sharma A (1998) PEO based solid polymer electrolyte plasticized by dibutyl phthalate. *Solid State Ionics* 113–115: 179–186.

Suthanthiraraj SA, Sheeba DJ, Paul BJ (2009) Impact of ethylene carbonate on ion transport characteristics of PVdF-AgCF$_3$SO$_3$ polymer electrolyte system. *Mater Res Bull* 44: 1534–1539.

Takahashi T (1989) Recent trends in high conductivity solid electrolytes and their applications: An overview. In: Laskar AL, Chandra S (eds) *Superionic Solids and Solid Electrolytes Recent Trends*. Academic Press, San Diego, pp. 1–36.

Tang S, Guo W, Fu Y (2021). Advances in composite polymer electrolytes for lithium batteries and beyond. *Adv Energy Mater* 11, 2000802.

Tarafdar S, De SK, Manna S, De U, Nanda P (2010) Variation in viscosity and ion conductivity of a polymer-salt complex exposed to gamma irradiation. *Pramana-J Phys* 74: 271–279.

Tominaga Y, Endo M (2013) Ion-conductive properties of polyether-based composite electrolytes filled with mesoporous silica, alumina and titania. *Electrochim Acta* 113: 361–365.

Tong Y, Chen L, Chen Y, He X (2012) Enhanced conductivity of novel star branched liquid crystalline copolymer based on poly(ethylene oxide) for solid polymer electrolytes. *Appl Surf Sci* 258: 10095–10103.

Vijayakumar G, Lee MJ, Song M, Jin S-H (2009) New liquid crystal-embedded PVdF-co-HFP-based polymer electrolytes for dye-sensitized solar cell applications. *Macromol Res* 17: 963–968.

Vioux A, Viau L, Volland S, Bideau JL (2010) Use of ionic liquid in sol-gel; ionogels and applications. *Comptes Rendus Chim* 13: 242–255.

Wang L, Yang W, Wang J, Evans DG (2009) New nanocomposite polymer electrolyte comprising nanosized ZnAl$_2$O$_4$ with a mesopore network and PEO-LiClO$_4$. *Solid State Ionics* 180: 392–397.

Watanabe M, Nagano S, Sanui K, Ogata N (1986) Ion conduction mechanism in network polymers from poly(ethylene oxide) and poly(propylene oxide) containing lithium perchlorate. *Solid State Ionics* 18&19: 338–342.

West AR (1999) Electrical properties. In: *Basic Solid State Chemistry*. John Wiley & Sons West Sussex, Vol. 2, pp. 293–372.

Wright VP (1975) Electrical conductivity in ionic complexes of poly(ethylene oxide). *Brit Polym J* 7: 319–327.

Wright VP (1998) Polymer electrolytes-the early days. *Electrochim Acta* 43: 137–1143.

Wu F, Feng T, Bai Y, Wu C, Ye L, Feng Z (2009) Preparation and characterization of solid polymer electrolytes based on PHEMO and PVDF-HFP. *Solid State Ionics* 180: 677–680.

Xi J, Qiu X, Zheng S, Tang, X. (2005) Nanocomposite polymer electrolyte comprising PEO/LiClO$_4$ and solid super acid: Effect of sulphated-zirconia on the crystallization kinetics of PEO. *Polym* 46: 5702–5706.

Yahya MZA, Arof AK (2003) Effect of oleic acid plasticizer on chitosan-lithium acetate solid polymer electrolytes. *Eur Polym J* 39: 897–902.

Yang C-C, Chien W-C, Li YJ (2010) Direct methanol fuel cell based on poly(vinyl alcohol)/ titanium oxide nanotubes/poly(styrene sulfonic acid) (PVA/nt-TiO$_2$/PSSA) composite polymer membrane. *J Power Sources* 195: 3407–3415.

Yang C-C, Chiu S-J, Lee K-T, Chien W-C, Lin C-T, Huang C-A (2008) Study of poly(vinyl alcohol)/titanium oxide composite polymer membranes and their application on alkaline direct alcohol fuel cell. *J Power Sources* 184: 44–51.

Yang C-C, Li YJ, Liou T-H (2011) Preparation of novel poly(vinyl alcohol)/SiO$_2$ nano-composite membranes by a sol-gel process and their application on alkaline DMFCs. *Desalination* 276: 366–372.

Yang C-C, Lin C-T, Chiu S-J (2008) Preparation of the PVA/HAP composite polymer membrane for alkaline DMFC application. *Desalination* 233: 137–146.

Yang H, Wu N (2022). Ionic conductivity and ion transport mechanisms of solid-state lithium-ion battery electrolytes: A review. *Energy Science and Engineering* 10, 1643–1671.

Yang Y, Guo X-Y, Zhao X-Z (2012) A novel composite polysaccharide/inorganic oxide electrolyte for high efficiency quasi-solid-state dye-sensitized solar cell. *Procedia Eng* 36: 13–18.

Yang Y, Zhou CH, Xu S, Hu H, Chen BL, Zhang J, Wu SJ, Liu W, Zhao XZ (2008) Improved stability of quasi-solid-state dye sensitized solar cell based on poly(ethylene oxide)-poly(vinylidene fluoride) polymer-blend electrolytes. *J Power Sources* 185: 1492–1498.

Yang, C.-M.; Kim, H.-S.; Na, B.-K.; Kum, K.-S.; Cho, B.W. Gel-type polymer electrolytes with different types of ceramic fillers and lithium salts for lithium-ion polymer batteries. *J Power Sources* 2006, 156, 74–580.

Ye Y-S, Rick J, Hwang B-J (2013) Ionic liquid polymer electrolytes. *J Mater Chem A* 1: 2719–2743.

Yim T, Kwon MS, Mun J, Lee KT (2015). Room temperature ionic liquid-based electrolytes as an alternative to carbonate-based electrolytes. *Israel J Chem* 55, 586–598.

Yuan F, Chen H-Z, Yang H-Y, Li H-Y, Wang M. (2005) PAN-PEO solid polymer electrolytes with high ionic conductivity. *Mater Chem Phys* 89: 390–394.

Zapata P, Lee J-H, Meredith JC (2012) Composite proton exchange membranes from zirconium-based solid acids and PVDF/acrylic polyelectrolyte blends. *J Appl Polym Sci* 124: E241–E250.

Zhang J, Huang X, Wei H, Fu J, Huang Y, Tang X (2010) Novel PEO-based solid composite polymer electrolytes with inorganic-organic hybrid polyphosphazene microspheres as fillers. *J Appl Electrochem* 40: 1475–1481.

Zhang P, Yang LC, Li LL, Ding ML, Wua YP, Holze R (2011) Enhanced electrochemical and mechanical properties of P(VDF-HFP)-based composite polymer electrolytes with SiO$_2$ nanowires. *J Membr Sci* 379 80–85.

6 Next-Generation Supercapacitors with Sustainably Processed Carbon Quantum Dots

Grishika Arora, Tejas Sharma, Foo Wah Low,
Chai Yan Ng, Pramod K. Singh,
Chiam-Wen Liew, and Hieng Kiat Jun

6.1 INTRODUCTION

The advanced version of batteries has become possible in recent years with the introduction of supercapacitors (SCs), which possess improved power density (PD) up to 10 KW/kg, energy density (ED), and specific capacitance value (Gupta et al., 2018; Hashmi et al., 2007). Based on the charge system, SCs can be divided into electric double-layer capacitors (EDLCs), pseudocapacitors, and hybrid capacitors.

EDLCs utilize carbon electrodes or their derivatives, which exhibit significantly higher capacitance due to the electrostatic double-layer formation rather than electrochemical pseudocapacitance. This phenomenon occurs at the interface between the conductive electrode surface and the electrolyte, leading to the separation of charges in the Helmholtz double layer (Shi et al., 2011). As such, the specific capacity of the device can be increased by increasing the area of the electrode surface, thus enhancing the conductivity and the specific capacitance of the electrode.

Metallic oxides like V_2O_5, Fe_3O_4, and RnO_2 are commonly used as active layer materials in pseudocapacitors. In comparison to EDLCs, pseudocapacitors exhibit higher energy storage capabilities and a specific level of capacitance. These are attributed to the occurrence of Faradaic charge transfer processes between the electrode and the electrolyte ions, which contribute to the overall capacitance of the pseudocapacitor system. However, due to the limitations of this Faradic reaction, pseudocapacitor electrodes show poor electrical conductivity, attributed to their inferior solidity (Xiao et al., 2009; Yang et al., 2013).

Carbon quantum dots (CQDs) have been gaining research interest in recent years. CQDs offer several advantages such as low toxicity, excellent solubility, strong and long-lasting photoluminescence (PL), exhibit high specific surface area, making them more commercially promising than their semiconductor counterparts (Lim et al., 2015; Hu et al., 2016). These advantages have received significant attention for CQDs in the field of material science. They also exhibit remarkable liquid dispersibility,

DOI: 10.1201/9781003314424-6

ease of functionalization, visual characteristics, and can be produced on a large scale at low cost. Due to these properties, CQDs have been employed in diverse sectors, encompassing cell molecular biology, drug delivery systems, carbon capture technologies, gas preservation applications, and cancer cell imaging (Ezugbe and Rathilal, 2020; Abbas et al., 2018; Ren et al., 2018).

As reported, there are several advantages to utilizing CQDs in biomedical applications compared to conventional semiconductor quantum dots and organic dyes (Wang et al., 2019). The extremely small size, remarkable biocompatibility, high biochemical inertness, and adjustable hydrophilic nature of CQDs make them widely used tools for scientific research such as biomedical and cancer studies (Ezugbe and Rathilal, 2020; Abbas et al., 2018; Li et al., 2017; Tuerhong et al., 2017). The small size of CQDs, particularly those with a particle size of less than 10 nm offers benefits such as good biocompatibility, low luminescence, enhanced molecule-membrane stability, biological inertness, favorable water mobility, and flexibility. These properties make CQDs highly sought after in research endeavors. CQDs outperform in organic dye-based systems that are dyes made from organic compounds and semiconductor quantum dots with metallic cores in biological systems, establishing their dominance in biosensor and bioimaging applications (Ren et al., 2018). Owing to their diminutive dimensions, CQDs demonstrate an expanded surface area and harness the quantum confinement effect (QCE), which manifests as discrete energy levels. This phenomenon yields heightened energy storage capabilities, enabling the EDLC to accommodate a greater charge. CQDs also can function as conductive additives, augmenting the conductivity of the electrode material. Furthermore, their environmentally benign nature contributes to the enhanced performance of the EDLC SC.

6.2   GENERAL CHARACTERISTICS OF SUPERCAPACITORS AND EDLCS

The efficacy of an SC hinges on factors such as the composition of its electrode and electrolyte components, and the mechanisms involved in charge storage (Iro, 2016; Braga et al., 2023). In response to challenges arising from the overreliance on non-renewable fossil fuels and the pursuit of cost-effective and environmentally friendly energy alternatives, the need for new and sustainable energy sources has become paramount. Despite this, existing energy storage solutions fall short of accommodating extensive storage demands. As a result, the demand for high-power and high-energy-density storage technologies is growing progressively (Iro, 2016). In general, SC technologies can be categorized into EDLC, hybrid capacitor, and pseudocapacitor (refer Figure 6.1).

SCs have emerged as energy storage devices that solve most of the problems associated with traditional batteries. Due to the quick electrostatic charge storage mechanism at the electrode–electrolyte interface, SCs like EDLCs are capable of being charged and discharged quickly, frequently in a matter of seconds. This makes them perfect for uses that call for brief bursts of electricity, such as regenerative braking in automobiles. Due to their high PD, EDLCs can deliver a considerable amount of energy in a brief length of time. The performance of EDLCs can withstand

Mechanisms

Types

Supercapacitors

Electrochemical double-layer capacitors (EDLC)

Ion adsorption/desorption by electrostatic forces at the electrode–electrolyte interface

Hybrid capacitors

Ion adsorption/desorption and reversible faradic reaction in an electrochemical double layer

Pseudocapacitors

Electrolyte ions undergo a reversible faradic redox reaction with electroactive materials

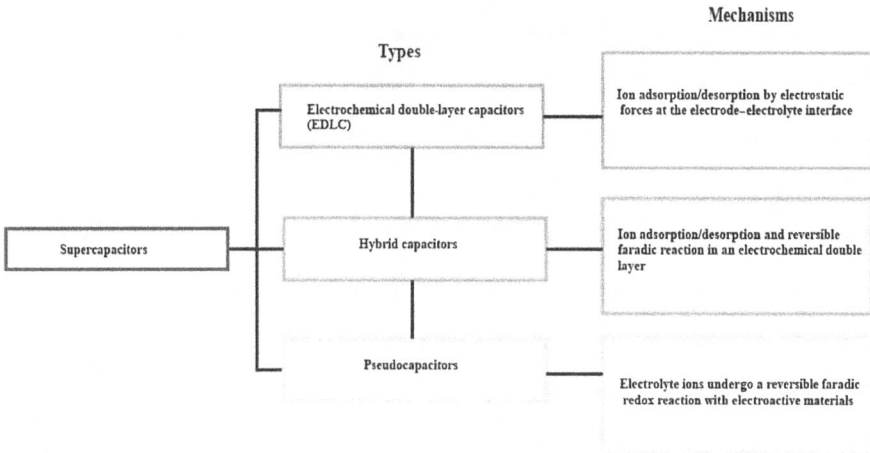

FIGURE 6.1   Varieties of supercapacitors and their energy storage mechanisms.

numerous charge and discharge cycles without noticeably degrading and EDLCs are safer, light, and environmentally friendly. SCs are found in a variety of applications, including electric buses and renewable energy systems. SCs are rated in Farads, with a nominal number of 500 Farad available (Braga et al., 2023).

EDLCs offer higher ED than ordinary capacitors due to their wide cross-sectional area and shorter lengths of the charge separation. As a result, they can achieve high capacitance values. When referring to EDLC devices, there are a lot of benefits and drawbacks. For instance, because of their non-Faradaic electrical mechanism, filling and discharging cycles are reversible. Additionally, this results in a high rate of charging and discharging, highly steady cycling capacity of at least 10^6 cycles or more, and minimal degradation. A fundamental disadvantage of EDLCs is the limited availability of electrode materials, where highly conductive electrodes are required in EDLC devices. Ionic conductive electrolytes, recently developed, can address this weakness. Because of this, the study of ionic electrolytes, especially sturdy ones, is currently popular research (De Klerk et al., 2018; Najib and Erdem, 2019).

6.3   WORKING PRINCIPLE AND CLASSIFICATION OF SUPERCAPACITORS

Electrostatic capacitors are the foundation of the SC operating system (refer Figure 6.2a). Equation 6.1 is the basic formula for all types of capacitors:

$$C = (\varepsilon_0 \times \varepsilon_r \times A)/d \qquad (6.1)$$

Electrode surface area, the interval within two electrodes, the dielectric material's relative permittivity, and the permittivity of empty space with opposing biases are all given in this equation by the letters $\varepsilon_0 \times \varepsilon_r$, A, and d, respectively.

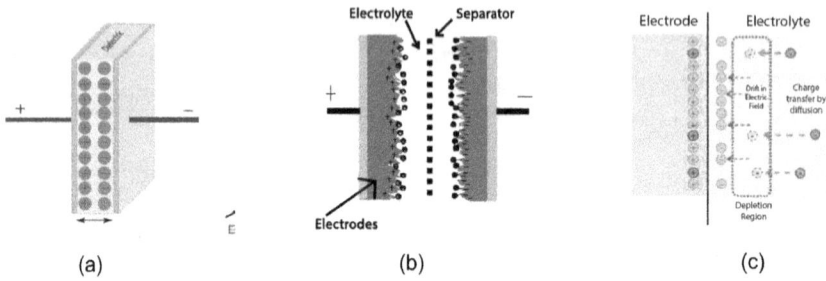

FIGURE 6.2 (a) The electrostatic capacitor's construction, (b) supercapacitor's structure, and (c) supercapacitor charge transfer between the diffusion and Helmholtz layers.

By modifying certain factors of a capacitor, such as the dielectric constant of the material used, the surface area of the electrodes, and the interplanar thickness, it is possible to increase the capacitance value according to Equation 6.1 (Najib and Erdem, 2019; Sahin et al., 2020). However, in the case of an SC, the structure differs from a traditional capacitor.

An SC's basic structure consists of current collectors, typically made of aluminum, as the electrodes. These electrodes are separated by a separator, which is usually made of a material that can be dissolved in an organic or aqueous electrolyte, as depicted in Figure 6.2b. The functioning of an SC relies on energy storage through the diffusion of ions from the electrolyte into the outermost layers of its two electrodes (Wang et al., 2017). The electrical double layer (EDL) forms at the two interfaces, leading to the generation of an EDLC.

When the electrodes and electrolytes come into contact, a thin layer is formed, known as the Helmholtz capacitance layer. At the electrode/electrolyte interface, the Helmholtz layer is assumed to move within an electric field in close proximity to the contact area. This movement of charge within the Helmholtz layer is what gives rise to the capacitance of an SC. The charge from the bulk electrolyte diffuses into the depletion zone created by this drift of charge in the Helmholtz layer (Wasterlain et al., 2006; Kuparowitz, 2010).

The Helmholtz theory utilizes two superficial charge distributions to illustrate the physical processes occurring at the interface between an ionic electrode and an electronic electrode. This layer plays a crucial role in the distribution of charge over time during the charging process as SCs excel in applications demanding swift energy transfer and elevated power performance due to their ability to effectively store and discharge electrical energy, as shown in Figure 6.2c.

SCs differ from ceramic or electrolyte capacitors in that they do not rely on a dielectric material. Instead, SCs utilize the formation of an EDL at the interface between a solid electrode, a carbon-based powder, and a liquid electrolyte. In order to prevent direct contact between the electrodes, a separator is used to keep them apart.

The electrodes in SCs are coated with activated carbon (AC) powder and placed on current collectors. During the charging process, the positive electrode accumulates ions and electrons, while the negative electrode accumulates vacancies.

This alignment of charge occurs at the contact interface between the electrodes. This arrangement forms the structure of an EDLC, as illustrated in Figure 6.3a. The discharge process, shown in Figure 6.3b, involves the dispersion of ions throughout the electrolyte, as they are not strongly attracted to the current collectors. Consequently, the charge states of both current collectors decrease (Thangaraj et al., 2019; Kumari et al., 2023; Najib and Erdem, 2019).

When considering the advantages and disadvantages of SCs, it is important to view them as complementary energy storage options alongside other technologies (Glavin and Hurley, 2012). SCs are voltage-regulated devices that require responsible usage to ensure that the voltage remains within specified limits. Capacitors with organic solvents typically have a nominal voltage range of 2.5–2.7 V, while standard SCs with aqueous electrolytes have a maximum voltage range of 2.1–2.3 V (Salanne, 2017). To achieve higher voltages, SC cells must be connected in series. The maximum capacitance of an SC can vary from 1 to 1,000 F, and for larger applications, parallel connections of cells are required to achieve the desired capacitance (Sahin et al., 2020).

In Figure 6.4, SCs could either act as batteries or capacitors as they have inter-crossed power/energy density values (Kim et al., 2015). Conventional capacitors have a higher power efficiency than the SC, despite the SC having a higher energy efficiency. Compared to SCs, batteries have a higher ED, but the PD of the SCs is higher. Galvanostatic charge-discharge (GCD) and cyclic voltammetry (CV) electrochemical techniques are two prominent ways for the purpose of characterizing material capacitance (Kim et al., 2015; Najib and Erdem, 2019). Equation 6.2 represents the specific ED:

$$ED = \frac{1}{2}Cs(\Delta V)^2 = \frac{1C}{2m}(\Delta V)^2 \qquad (6.2)$$

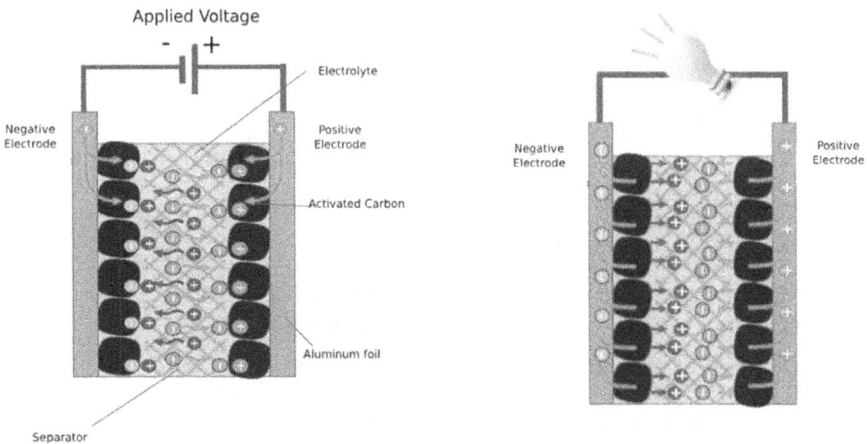

Applied Voltage

FIGURE 6.3 (a) Charging phase of supercapacitor and (b) discharging phase of supercapacitor.

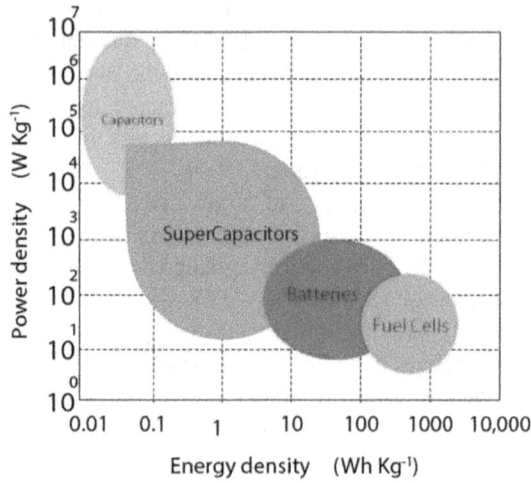

FIGURE 6.4 Power and energy densities of storage systems are compared.

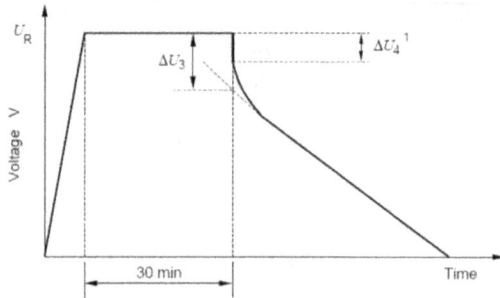

FIGURE 6.5 Current charging and discharging.

where V is the operating potential range and Cs is the specific capacitance from CV or GCD methods. A specific PD is the speed at which a device may transmit energy to outside loads while maintaining a consistent current density. Equation 6.3 represents the equation to determine the maximum specific PD, PD:

$$PD_{max} = \frac{(V)^2}{4mR_{ESR}} \tag{6.3}$$

where V is the potential range, m is the mass of the active elements, and R_{ESR} is the equivalent series resistance (ESR) within the cell (Kim et al., 2015).

There are techniques available to measure the ESR and capacitance of SCs through direct current (DC) characterization using steady current (Gualous et al., 2016). In these techniques, the SC is initially charged at a constant current (CC) until it reaches a very low voltage threshold. Then, it is connected to a DC source with a predefined voltage. During the discharge cycle, the SC is discharged at a CC, as shown in Figure 6.5 (Lawu et al., 2017; Kraev et al., 2016). For DC

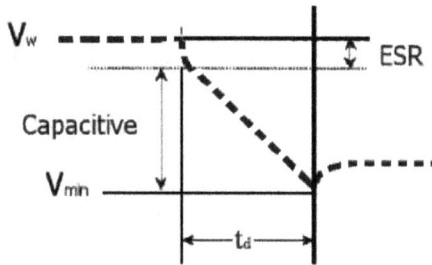

FIGURE 6.6 Supercapacitor discharging.

characterization, ESR and capacitance measurement techniques have been developed based on the charging and discharging of SCs at a CC. In this approach, the SC is initially charged using a CC until it reaches a specific voltage threshold. Subsequently, it is connected to a fixed-voltage DC source. The discharge process involves discharging the SC at a CC until an initial voltage drop is observed. By adjusting certain parameters, such as charging time, the duration of the charging process can be modified. The discharge profile, illustrated in Figure 6.6, provides insights into the discharging behavior (Glavin and Hurley, 2012; Prasad et al., 2019).

The voltage across the SC comprises both capacitive and resistive elements. The capacitive component accounts for voltage changes during the charging and discharging phases. On the other hand, the resistive component contributes to voltage variations due to internal resistance, which is quantified by the ESR.

The reactive component in the measurement represents the voltage fluctuations caused by the charge and discharge processes. Both the ESR and the resistive component provide insights into the voltage shift caused by the internal resistance of the SC. Unlike self-discharge, the redistribution of charge in SC leads to a relatively rapid voltage drop or increase within a short period of time (Prasad et al., 2019).

6.3.1 Electric Double-Layer Capacitors (EDLCs)

When it comes to energy storage, EDLCs, a specific subset of SCs, primarily utilize electric double-layer capacitance. In contrast, other types of SCs may employ additional mechanisms such as pseudocapacitance. The diverse designs of SCs result in a range of energy storage capabilities and characteristics, and these distinctions can be attributed to the variations in underlying mechanisms. EDLCs consist of an electrolyte, two electrodes made of carbon-based materials, and a separator. EDLC can keep the charges either electrostatically or by a non-Faradic mechanism that eliminates the requirement for charge transfers between the electrode and the electrolyte (Kiamahalleh et al., 2012). EDLCs utilize the idea of electrochemical double layers for energy storage. When the opposing charge attraction resulting in potential difference, the voltage is induced and there is no build-up of charges on the electrode surface. Instead, electrolyte ions are spread to the charged-opposing electrode pores and over the separator as shown in Figure 6.7a. An additional charge layer is formed

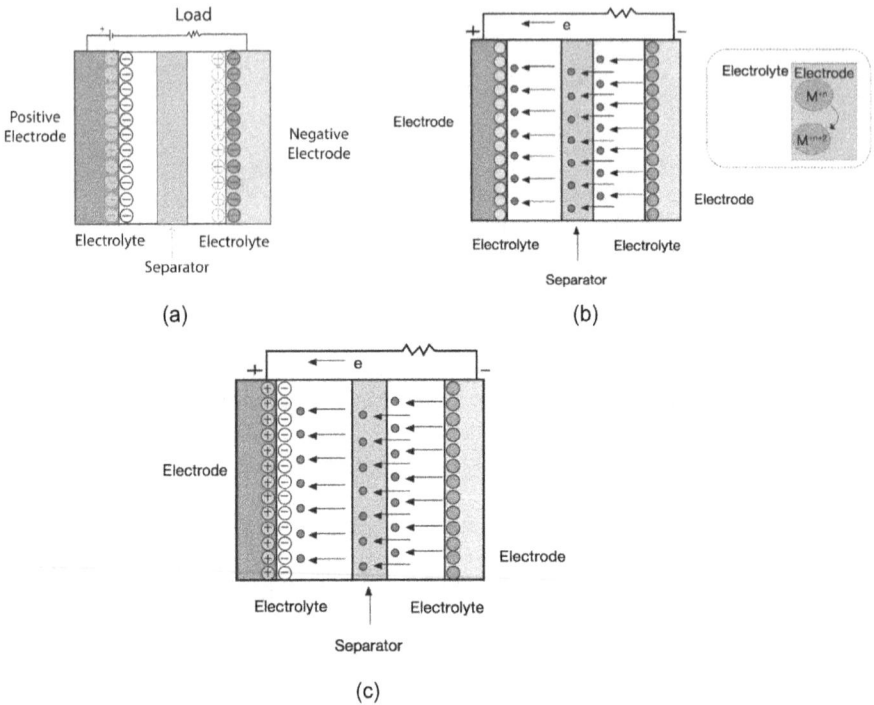

FIGURE 6.7 Illustration of several SC types for energy storage: (a) EDLCs, (b) pseudoca-pacitors, and (c) hybrid capacitors. Electrochemical capacitors or EDLCs, commonly referred to as supercapacitors, Boost Caps, or Gold Caps, have high capacitance and low internal resistance. Therefore, perfect for a wide range of applications.

to prevent ion recombination within the electrodes. The presence of a double layer in EDLCs facilitates a high ED, enhances the specific surface area, and reduces the electrode separation (Jalal et al., 2021; Choi and Yoon, 2015).

Furthermore, EDLCs' storage method enables rapid absorbance of energy, dis-tribution, and efficient power generation. The non-Faradic phase, characterized by the absence of chemical changes, allows EDLCs to endure millions of cycles, whereas batteries are typically limited to just a few thousand cycles at best. This distinction could explain some of the disparities between batteries and EDLCs. Additionally, when high potential cathodes or graphite anodes are used in Li-ion batteries, the solid electrolyte interphase is formed. However, there is no need for an electrolyte solvent during charging (Jalal et al., 2021; Iro, 2016; Simon and Gogotsi, 2008).

6.3.2 PSEUDOCAPACITORS

This type of SC is characterized by employing high electrochemical pseudoca-pacitance materials and the electrodes are made from metal oxides or conducting polymers. The charge storage mechanism involves Faradaic processes, such as oxidation–reduction reactions, which include charge transfer between the electrode and the electrolyte as shown in Figure 6.7b (Jalal et al., 2021).

Applying an electric field to a pseudocapacitor triggers reduction and oxidation on the electrode material due to charge passage through two layers, resulting in the Faradic current across an SC cell. Additionally, compared to EDLCs, pseudocapacitors can attain high specific capacitance and energy densities, thanks to the Faradic mechanism (Jalal et al., 2021; Chen et al., 2014). In the early days, efforts have been made to develop novel reaction processes for efficient and ultrafast charging/discharging by utilizing rechargeable batteries and SCs in the creation of electrode materials such as conducting polymers and metal oxides. In fact, it must be extremely energetic SCs and also have high power and long cycle life. Through the physical manipulation of electrode materials, researchers have discovered an intercalation pseudocapacitance contribution, akin to the pseudocapacitance found in certain electrode materials for metal-ion batteries (Wang et al., 2017).

6.3.3 HYBRID CAPACITORS

In a particular type of SC, the electrodes are asymmetric, with one primarily showcasing electrostatic capacitance, while the other demonstrates the transmission of chemical capacitance signals. Capacitors that are in hybrid mode have combined performance characteristics that were previously impossible. Additionally, they are fusing the most advantageous aspects of pseudocapacitors and EDLCs into an individual SC (Jalal et al., 2021). Although hybrid capacitors have not garnered as much attention as pseudocapacitors or EDLCs, ongoing efforts are dedicated to enhancing the hybrid capacitor technology and establishing more precise quantitative models for their performance.

Hybrid capacitors have surpassed EDLCs as the primary SCs class due to their enormous simplicity in tuning their performance and design, as well as the increased focus on developing high energy and high cycle life SCs (Sharma and Kumar, 2020). The mechanism for storing charge is shown in Figure 6.7c. The charge storage mechanism in hybrid capacitors is a synergy of electrostatic double-layer capacitance and pseudocapacitance. The EDLC mechanism operates at the electrode–electrolyte interface, where electrical charges accumulate due to ion separation within the electrolyte. This leads to the formation of a dual charge layer and, consequently, elevated capacitance. In contrast, the pseudocapacitance mechanism involves redox reactions within the electrode material itself, resulting in reversible Faradaic processes that contribute to additional charge storage. By combining these two mechanisms, hybrid capacitors exhibit higher energy density than standard EDLC capacitors, making them suitable for diverse energy storage applications. Asymmetric, composite, and battery-type hybrid capacitors are categorized into these three groups, which could be distinguished by their electrode configuration.

By incorporating the polymerized carbon-based materials conduction or metal oxides in a single electrode via composite electrodes, electrodes will have both chemical and physical processes for storing energy. Binary and ternary composites are the two primary types. Binary composites are electrodes made of only two components; ternary composites are electrodes built of three different materials

(Jalal et al., 2021). Through the integration of an EDLC with asymmetric hybrid capacitors, a combination of Faradic and non-Faradic processes is harnessed to form a pseudocapacitor electrode. In this approach, the carbon material functions as the negative electrode while the metal oxide or conducting polymer acts as the positive electrode (Jalal et al., 2021).

On the other hand, batteries are a unique fusion of an SC electrode with a battery electrode. By fusing SC recharge times and battery qualities to obtain both battery and SC properties in a single cell, this approach highlights the need for greater power batteries and high energy capacitors. Battery-powered hybrids haven't been the subject of many investigations though (Jalal et al., 2021; Costa et al., 2022).

6.4   CARBON QUANTUM DOTS (CQDS)

When combining sp^2 and sp^3 carbons, CQDs often have a quasi-spherical crystalline core with a surface coating of chemical functional groups as shown in Figure 6.8. Most of the carboxyl groups on the surface are both biocompatible and highly hydrophilic (Liu et al., 2021).

CQD synthesis has been accomplished using a variety of demonstrated techniques. The primary goal of a lot of research is to identify the most effective, environmentally friendly, and high-yielding synthetic approach (Magesh et al., 2022). Top-down and bottom-up techniques are the two broad categories used to categorize the routes for CQD synthesis (Gonzalez et al., 2016). Using top-down method techniques like electrochemical oxidation, chemical oxidation, and ultrasonic technologies, bigger carbonaceous materials and structures are broken down or divided into smaller pieces (Lim et al., 2015; Wang et al., 2019). In contrast, the bottom-up approach involves the formation of small CQDs by structuring carbon materials to achieve the desired size. Typical bottom-up methods for CQD creation include solvothermal, hydrothermal, ultrasonic, thermal decomposition, pyrolysis, and carbonization processes (Figure 6.9) (Desmond et al., 2021; Costa et al, 2022).

For the effective creation of CQDs, several top-down techniques have been introduced, such as etching with bigger carbon materials, laser treatment, thermal

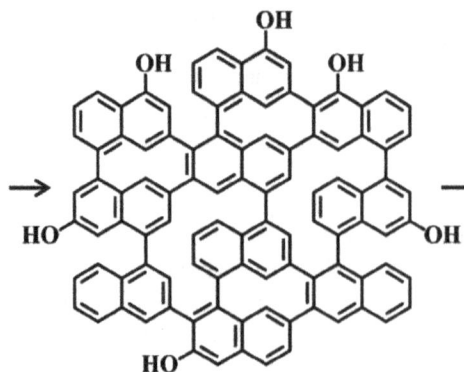

FIGURE 6.8   The general chemical structure of CQDs. Adapted from Desmond et al. (2021).

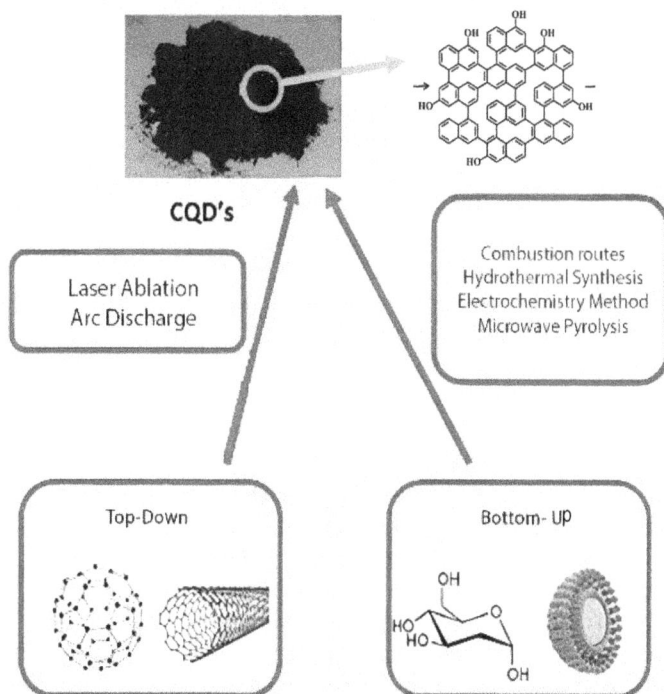

FIGURE 6.9 A general diagram demonstrating the creation of CQDs from a number of efficient top-down and bottom-up techniques. Adapted from Wang et al. (2019). The common beginning supplies for the techniques presented in the bundles are displayed in the top-down and bottom-up approaches. The completed CQD product and its general chemical composition are displayed at the top of the illustration.

oxidation, electrolytic oxidation, hot injection, distillation, and microwave irradiation. Bottom-up procedures typically produce consistent-sized particles that are preferred for utilization in biological applications whereas top-down approaches typically produce tiny flakes or particles that come in a variety of sizes (Desmond et al., 2021; Tuerhong et al., 2017).

6.4.1 PROPERTIES OF CQDS

6.4.1.1 Absorbance

The typical characteristic of the absorbance of CQDs is a broad and strong absorption peak in the UV-visible spectrum. Overall, it is anticipated that the reflected light peaks of CQDs in the UV-visible range are expected to be π-π* transition of sp^2 conjugated carbon and n-π* transition and heteroatoms hybridization such as N, S, P, and so on. Passivation of the surface or other types of modification can change the absorption characteristic (Sadat et al., 2019). Jiang et al. used three isomers of phenylenediamines to establish a simple hydrothermal method for making red, green, and blue luminous CQDs (Jiang et al., 2015). The obtained CQDs' UV-visible absorption spectra showed an analogous pattern. It is interesting to note that three CQD absorption transitions were red-shifted, revealing lower electronic spectrum gaps than those of their equivalent antecedents (Wang et al., 2019).

6.4.1.2   Photoluminescence (PL)

PL refers to the emission of light when the small carbon structures are stimulated by light. As CQDs absorb photons, they undergo electronic transitions, releasing energy in the form of visible light upon returning to their original state. PL represents one of the most intriguing properties of CQDs, both in terms of scientific study as well as real-world applications (Li et al., 2012). The CQDs' significant PL dependence on both the emission wavelength and intensity remains consistent throughout. This unusual phenomenon might be caused by the surface-based optical selection of CQDs with varied emissive traps or nanoparticles with different sizes (Zhang et al., 2017). When the emission characteristics of CQDs were examined under 470 nm radiation with different levels, the varied CQDs particle size and PL emission may be seen in the broad and oscillation frequency-dependent PL emission spectrum that was observed. It was discovered that as the concentration of the CQDs solution grew, the PL strength first rose and then fell (Wang et al., 2019; Zhang et al., 2017). Depending on the unique properties of the CQDs and the wavelength of excitation, the usual PL spectra show a broad emission peak in the visible range, often spanning from blue to green.

6.4.1.3   Electroluminescence

Electroluminescence (ECL) pertains to the phenomenon where the carbon nanostructures emit light when subjected to an electric field. This emission occurs due to electron energy level transitions triggered by the passage of an electric current through CQD-based materials. Since semiconductor nanocrystals are known to exhibit ECL, it should come as no surprise that CQDs have sparked a variety of research interests in ECL that can be useful in electrochemical domains as disclosed in a light-emitting diode device based on CQDs, where the driving current may be used to regulate the emission color (Sk et al., 2014). Under various working voltages, the same CQDs showed color-switchable ECL that ranged from blue to white.

Two concepts were developed based on the emission from the band gap of the conjugated "p'" domain and the reduction effect arising from surface issues. These concepts were developed by the researchers to explain the luminescence process of CQDs (Yuan et al., 2015). The QCE of p-conjugated electrons in the sp^2 atomic framework is what gives CQDs' conjugated p domain fluorescence emission its PL features. These characteristics can be changed by varying the size, edge configuration, and shape of the CQDs. Among other surface faults in CQDs, sp^2 and sp^3 hybridized carbon produce fluorescence emission in these materials. This defect also depends on peak position and fluorescence intensity (Wang et al., 2019; Chen et al., 2021).

6.4.2 SYNTHESIS OF CQDS

6.4.2.1 Bottom-Up Approaches

6.4.2.1.1 Hydrothermal/Solvothermal Method

The process of hydrothermal carbonization (HTC), which is a way to create new carbon-based compounds from a variety of precursors, is cost-effective, safe, and nontoxic. Generally, a sealed organic precursor solution is heated to a high temperature in a hydrothermal reactor (Wang and Hu, 2014). Small organic molecules are suspended with the reaction precursor in water or an organic solvent. It is then put into the Teflon-lined stainless-steel autoclave, which is frequently utilized in laboratories for the hydrothermal process (He et al., 2018). Organic molecules are then combined at a reasonably high temperature to create carbon-seeded cores that grow into 10 nm or smaller CQDs (Azam et al., 2021).

Given its uncomplicated synthesis procedure and adaptable hetero-atom doping, this technique offers a promising avenue for producing and synthesizing novel catalysts tailored for energy applications, offering the flexibility to customize doping ratios and electronic structures. According to Atchudan et al., the resulting CQDs have a near-spherical morphology with an average diameter of 5 nm and an adequate level of graphitization (Magesh et al., 2022; Atchudan et al., 2021). Because of their functionality, the water solubility of CQDs is very good and exhibits bright fluorescence without further surface treatment, with an acceptable quantum yield (QY) of 20%. CQDs have high biocompatibility and fluorescence stability, making them ideal candidates in bioimaging applications. To enhance water solubility and facilitate binding to specific cell receptors, Mahani et al. employed a modified approach involving transferrin (TF) and HTC of citric acid (Magesh et al., 2022; Mahani et al., 2021). For the purpose of delivering the drug (doxorubicin) to breast cancer cells, the drug was applied onto a carrier composed of TF-CQDs (MCF-7). The results also indicate the potential utility of Dox-loaded TF-CQDs for precise drug delivery.

When evaluating the benefits of the strategies employed in CQD production, the hydrothermal method stands out as a straightforward approach. It yields CQDs with high QY and controlled particle size, thereby reducing the need for expensive chemicals. Additionally, CQDs produced through hydrothermal methods find extensive applications in the fabrication of sensors, SCs, solar cells, and drug delivery systems (Magesh et al., 2022).

6.4.2.1.2 Acidic Oxidation

A commonly used method involves exfoliating and deconstructing hydrophilic functional groups, like hydroxyl or carboxyl groups, which are incorporated into the structure to convert bulk carbon into nanoparticles (Shen and Xia, 2014). It was reported that acidic oxidation and hydrothermal reduction were used to create hetero-atom-doped CQDs on a massive scale. To begin, carbon nanoparticles made using Chinese ink were oxidized using a mixture of H_2SO_4, HNO_3, and $NaClO_3$ (Magesh et al., 2022). Then, nitrogen, sulfur, and silicon sources were hydrothermally reacted with the oxidized CQDs. Di-methyl formamide (DMF), sodium hydrosulfide (NaHS), and sodium selenide (NaHSe) have all been used as precursors. The produced N-CQDs, S-CQDs, and Se-CQDs had higher QY, greater PL performance, and longer adjustable fluorescence lifetimes when compared to pure CQDs. The results showed that the heavily doped heteroatoms can modify the PL properties that are advantageously related to the electro-negativities of N, S, and Se. The active heteroatoms on the surface would modify the electronic structure of the associated CQDs when used as electrocatalysts, resulting in good electrocatalytic activity (Magesh et al., 2022; Feng and Zhang, 2019). Using this technique, Feng et al. produced CQDs from coke, which revealed 9.2% of QY. After a one-step chemical oxidation utilizing coke as the starting material and hydrogen peroxide as the oxidizing agent, the product had a blue-green-red spectral makeup of up to 48%. This study demonstrated the intriguing uses of coke-based CQDs as possible phosphors in light-emitting devices (Monte-Filho et al., 2019).

The acidic oxidation process is a quick and efficient way to make CQDs on a wide scale. However, the use of some potent oxidizers in this procedure could result in burns or explosions, and the post-processing is also exceedingly challenging (Magesh et al., 2022).

6.4.2.1.3 Microwave Pyrolysis

Because of its flexibility of use and commercial viability, microwave pyrolysis is a well-known bottom-up approach (refer Figure 6.10) (Feng and Zhang, 2019). Microwave irradiation induces atomic-level heating through the interaction of polar solvent molecules with electric and magnetic fields, resulting in rapid and energy-efficient reactions without physical contact or intermediaries. Compared to existing methods, this approach offers efficient CQD synthesis, rapid processing, eco-friendliness, energy/time efficiency, low impurity levels, controlled size/temperature, consistency, and precise reaction parameter control (Magesh et al., 2022; Tungare et al., 2020).

To create extremely luminous CQDs, Liu et al. discussed about microwave-mediated pyrolysis of citric acid with different amine compounds. The simultaneous roles that the amine molecules, particularly primary amine molecules, play in the N-doping process and surface passivation of the CQDs improve the PL performance. For CQDs made from citric acid and 1,2-ethylenediamine, the QY values significantly increased as the N content increased, exhibiting a QY up to 30.2%.

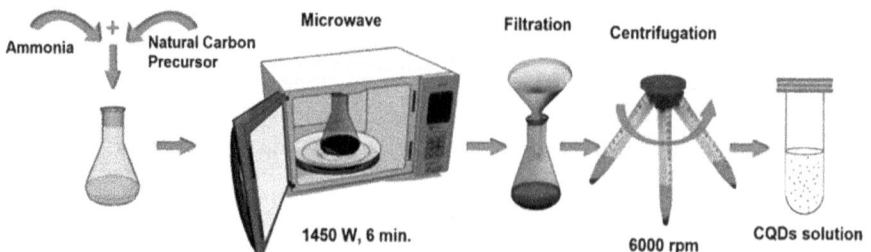

FIGURE 6.10 Synthesis method for CQDs using microwave-assisted pyrolysis. Adapted from Feng and Zhang (2019).

The resulting CQDs have significant potential for biomedical applications and are extremely biocompatible (Zhai et al., 2012).

With this technique, the synthesized CQDs have good physiochemical and optical characteristics, such as tunable emission, wide spectrum absorption, high charge carrier extraction, and high extraction of charge carriers. Although it is an easy one-stop method and is good for the environment, there is no way to regulate the size of the CQDs (Magesh et al., 2022).

6.4.2.2 Top-Down Approaches

6.4.2.2.1 Laser Ablation Method

In this method, the surface of the sample is illuminated by a powerful laser pulse. It then enters a thermodynamic condition with a high temperature (>900°C) and pressure (75 kPa), which rapidly warms up to a liquid phase where vapor crystallizes to produce carbon nanomaterials (Magesh et al., 2022; Hu et al., 2011). Radiating a precursor to carbon distributed in several typical organic solvents with a laser is an easy way to create CQDs (Figure 6.11). The synthesized carbon nanoparticles are aggregated with different-sized particles and lack any clear physical PL. After treating the sample with a weak HNO_3 solution, it is refluxed for 12 hours before polyethylene glycol is added to react with it (Zuo et al., 2015).

It was reported that the CQDs have adjustable and visible PL (Li et al., 2011). Modifying the PL properties of freshly synthesized CQDs and altered CQDs with strong PL and a diameter of 5 nm can alter the surface characteristics of CQDs. Laser ablation is a potent method for producing CQDs characterized by a narrow size distribution, excellent water solubility, and strong fluorescence (Magesh et al., 2022; Hu et al., 2009). When stimulated at 400 nm, CQDs produced fluorescence QYs that varied from 4% to 10% (Atchudan et al., 2021).

Typically, according to the solvent and precursor, CQDs produce more PL; hence, this method results in increased PL emissions (Cui et al., 2020). The shortcomings of this strategy are the huge number of raw materials needed and the inability to control particle sizes, which led to low QY (Magesh et al., 2022). It is

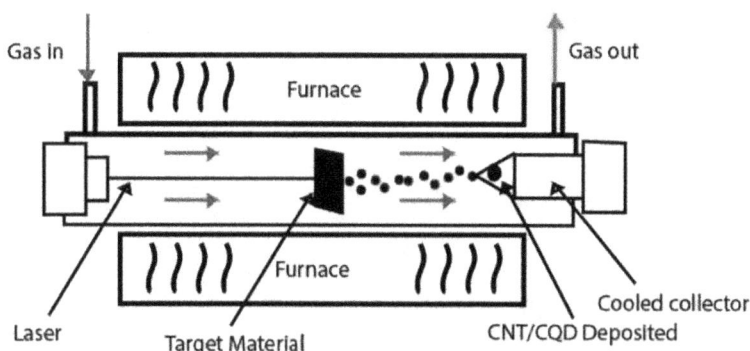

FIGURE 6.11 Diagram showing the development of CNT/CQD creation through laser ablation.

important to emphasize the speed and effectiveness of this technology as well as the customizable surface state of CQDs.

6.4.2.2.2 Arc-Discharge Method

Using gas plasma created in a sealed reactor inside of an anodic electrode, this technique breaks down the bulk carbon sources while reordering the carbon molecules within them (Figure 6.12) (Arora and Sharma, 2014). Under the influence of the electric current, the reactor's temperature can reach 4,000 K, producing plasma with enormous energy. CQDs are created in the cathode by the accumulation of carbon vapor. Different forms of carbon nanoparticles with various molecular weights and luminescence can be when synthesizing single-walled carbon nanotubes using the arc-discharge method (Dey et al., 2014). Under 365 nm excitation, the initially synthesized CQDs exhibit emissions in shades of orange, blue-green, and yellow. Through the introduction of the carboxyl group (hydrophilic) onto the CQDs' surface using HNO_3, enhanced water solubility is achieved. However, the resultant CQDs often possess a larger particle size of approximately 18 nm. This outcome can be attributed to the diverse scales and low yield of carbon particles generated during the arc-discharge method. Moreover, nanoparticles produced via the arc-discharge route may encompass intricate constituents that pose challenges to filtration and elimination processes (Zuo et al., 2015).

With this method, it is possible to create CQDs for bioimaging applications. However, the synthesized CQDs exhibit a limited QY and lack of particle size scalability. Nevertheless, they show a strong PL property (Magesh et al., 2022). Among all the approaches, the laser ablation approach is suggested as one of the top-down techniques for making CQDs since it allows for exact control over particle size, a reduction in contaminants, and the possibility to create CQDs with specific qualities. By employing laser irradiation to disintegrate a carbon target, this technique produces CQDs with the appropriate properties. Table 6.1 summarizes the unique features of the top-down and bottom-up approaches in synthesizing CQDs.

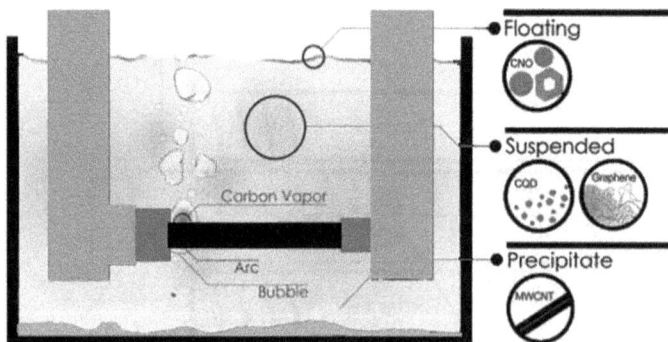

FIGURE 6.12   The graphic depicts the immersed arc-discharge in water method for creating CQDs.

TABLE 6.1

According to Reports, CQDs Have Been Synthesized Using Various Top-Down and Bottom-Up Approaches from a Diverse Range of Carbonaceous Starting Materials

Method (Top-Down/Bottom-Up)	Course of Action	Pros	Cons
Hydrothermal/solvothermal method	To create CQD, in water or an organic solvent, tiny organic molecules or polymers are dissolved and then transferred to an autoclave made of stainless steel lined with Teflon.	The CQDs' QY can reach 80%, and there are numerous carbon sources that can be used in the process.	Minimal QY. Gas is a byproduct of the reactor that might cause losses. inadequate control of size.
Acidic oxidation	After being oxidized by a solution of H_2SO_4, $NaClO_3$, and HNO_3 carbon nanoparticles are DMF, NaHS, and NaHSe hydrothermally interacted with each other.	Created CQDs outperform other CQDs in terms of prolonged fluorescence lifetime, high QY, and tunable PL performance.	Strongly doped heteroatoms may be caused by the electronegativity of S, N, and Se to have an impact on the PL characteristics.
Microwave pyrolysis	Forming a transparent solution by putting a carbon source and a sugar solution in water and heating it in a microwave in an inert environment.	A straightforward, ecologically harmless, commercially viable, and quick way to synthesis.	The highest QY reported is 34%, which is less than that for other methods in the table.
Laser ablation method	Laser irradiation is used to create the desired effect, toluene serves as the carbon source. Laser weapons are used to regulate CQD size.	Changing parameters like the irradiation period will allow you to easily modify the size and PL characteristics of CQDs.	This approach produces a modest yield of CQDs (4%–10%), making it non-green. Low QY, inadequate size control, adjustment required.
Arc-discharge method	An arc of current and voltage is created between two graphite rods that are employed as the anode and cathode electrodes, respectively, in an open vessel.	This procedure generates MWCNTs, which, in theory, might be further broken down to generate CQDs.	The soot from nanotubes contains a number of contaminants. Compared to nanotubes, the graphitic sheets in soot have stronger oxidative stability.

6.4.3 Synthesis of CQDs from Biomass Wastes

Despite the widespread availability at little cost, waste biomass generated from food and animal waste, forest byproducts, agricultural residues, and other sources remains largely unnoticed and underutilized as a potential raw resource. The entire synthesis process is specifically engineered to transform waste into an affordable and renewable final product, which holds promise for large-scale commercial production. Notably, no chemicals are employed during the preparation of CQDs from biomass (Thangaraj et al., 2019).

Among the biomass wastes, spent coffee grounds (CGs) are a prevalent organic waste byproduct known for their elevated carbon content. Studies have confirmed that the carbon obtained from spent CG exhibits exceptional characteristics, such as a high surface area, a well-developed graphitic structure, and a hierarchical porous arrangement, all of which are essential for promoting efficient capacitive behavior (Kumari et al., 2023). In comparison to heat drying, the hydrothermal approach utilized to produce CQD has been proven to be a more cost-effective method. The following sections discuss on practical methods of fabricating CQDs from biomass waste of spent CG.

6.4.3.1 Hydrothermal Method

In order to create fluorescent CQDs from the spent coffee beans as a carbon source, HTC is chosen. According to the work by Costa et al., the coffee beans were first cleaned in deionized water, air dried, and made in the powdered form (Costa et al., 2022). A stainless-steel autoclave with a 100 mL inner volume and Teflon coating was filled with some powdered coffee beans and heated at a continuous temperature of 200°C for 4 hours. It was then given time to cool down to room temperature. To get the clear, brownish solution, a 0.45 m membrane filter was used to filter out any undesirable big particles and unreacted organic molecules. The purified fluorescent CQD was then collected, and it was kept at 5°C for additional characterization and testing (Costa et al., 2022).

Carbon dots (CDs) were synthesized through a process involving the dissolution of CGs (approximately 5.0 g) in 100 mL of water (Chen et al., 2021). An amine additive was introduced at specific temperatures and durations within a sealed container. Following cooling to room temperature, the mixture was subjected to filtration using a cellulose membrane. The resulting brown filtrate was passed through syringe filters. Subsequently, measurements were conducted to determine insoluble residue concentrations, organic substance concentrations, and the quantity of CDs in aqueous dispersions. These measurements were executed in accordance with the previously described methods, wherein proportional volumes of organic solvents were employed for extraction.

In a separate study, Chen et al. concluded the CQDs produced from 3.125 g/L of CGs have the best fluorescence intensity, as determined by contrasting the fluorescence intensities of the three CQD concentrations under the identical circumstances, as shown in Figure 6.13a, and while investigating the reaction time, CQDs' fluorescence intensity in 6 hours was comparable to that in 12 hours, but it was much higher in 10 hours. Consequently, 10 hours is the ideal reaction time as shown in

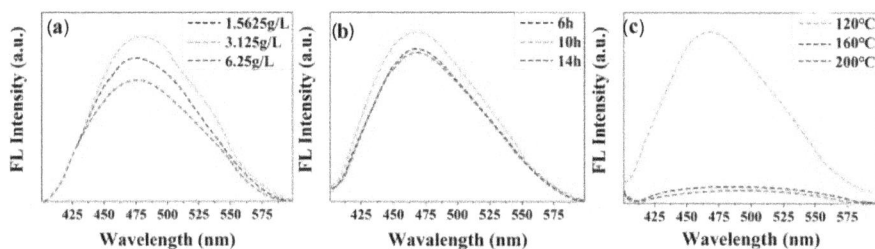

FIGURE 6.13   The ideal reaction conditions' emission spectra: (a) concentration, (b) time, and (c) temperature of the reaction. Adapted from Chen et al. (2021).

Figure 6.13b. While investigating the reaction temperature at 120°C, the intensity of this system peaked and remained constant. Additionally, at reaction temperature of 120°C, CQD fluorescence intensity is significantly higher than that at 160°C and 200°C as shown in Figure 6.13c (Chen et al., 2021).

6.4.3.2   Microwave-Assisted Hydrothermal Carbonization (Mw-HTC) Method

Coffee beans, a plentiful and environmentally unfavorable heavy trash, were found to be an appropriate source of vegetable biomass for the quick and efficient HTC technique that produces fluorescent CQDs. According to Shen et al., in order to make highly luminescent CQDs in the best chemical yields possible, an expanded criteria for the HTC and Mw-HTC's reactions were explored (Shen and Xia, 2014). Good reaction yields of CQDs with blue luminescence (up to 18%) were produced under carefully designed conditions. In conclusion, this research revealed CQDs from used CGs show outstanding toxic nitroaniline detecting capabilities for the initial period.

The CQDs made from CGs had outstanding fluorescence characteristics, with emission maxima that varied depending on an excitation wavelength (340–420 nm). Maximum fluorescence emission that shifts with decreasing excitation energies are typically linked to the presence of various particle sizes and/or surface states, which results in a distribution of visual gaps on the array (Figure 6.14). After 380 nm excitation, the maximum emission (HTC) is produced at 461 nm (Shen and Xia, 2014).

In the work by Arumugham et al., they used the Catharanthus roseus (a white blooming plant) leaves that were hydrothermally carbonized to create an eco-friendly fluorescence CQD probe, without the need of chemicals or laborious post-processing (Arumugham et al., 2020). This method was successfully proven. At the 330 nm excitation wavelength, they demonstrated a maximum QY of 28.2% providing stable photostability and an intriguing pH-dependent behavior of the produced CQDs. Studies on bioimaging has shown its capacity to diffuse across MCF-7 cells without any fluorescence emission degradation. These intriguing findings lead to the conclusion that Catharanthus roseus leaf-derived CQD may be a novel type of fluorescence probe with considerable possibilities in the study of cancer treatment and the reduction of environmental pollution (Arumugham et al., 2020).

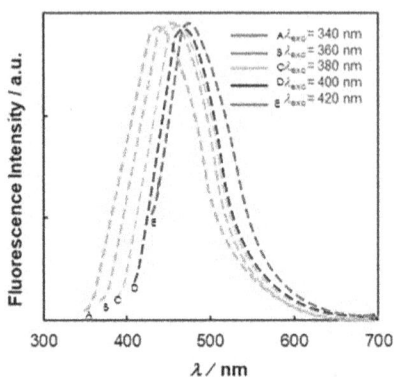

FIGURE 6.14 C-dot (HTC) aqueous dispersions (0.1 mg/mL) with standardized emission spectra derived from coffee grounds (excitation in the 340–420 nm region). Adapted from Shen and Xia (2014).

Utilizing electrode materials based on CQDs is widespread due to their economic viability and availability. Additionally, the flexibility to alter and control factors like pore configuration, surface traits, and surface area offers an added benefit when incorporating carbon materials into energy storage apparatus. In SCs, the storage mechanism hinges on the electrostatic allure of electrolyte ions to the surface of the electrode material. Thus, the presence of appropriately sized pores that align with the ions' dimensions becomes essential. Here are a few recent research endeavors aimed at enhancing SC performance.

6.5   APPLICATION OF "GREEN" CQDS IN SUPERCAPACITORS

There are number of works reported on CQDs applied in SC in recent years. A few notable works are discussed here. In the work by Inayat et al., they employed a hydrothermal approach to synthesize CQDs using tea leaves as a precursor (Inayat et al., 2023). These CQDs were intended for use as electrode materials in EDLC SCs. CV and GCD measurements were performed over a voltage range of −0.2 to +0.8 V compared to the Standard Calomel Electrode (SCE) reference electrode. These measurements were carried out using various sweep rates in a 1M H_2SO_4 electrolyte solution. Additionally, electrochemical impedance spectroscopy (EIS) experiments were conducted in a frequency range spanning from 0.01 to 10^5 Hz. The CQDs were prepared with a crystalline arrangement, exhibiting an average diameter of approximately 2–5 nm. Evaluations of the electrochemical performance of the CQDs revealed exceptional stability when employed as an SC electrode.

The CQD-based SC demonstrated remarkable characteristics, including a high specific capacitance of 302.0 F/g at a current of 0.5 A/g, impressive cyclability with a capacitance of 144.4 F/g at 20 A/g after undergoing 5,000 cycles, and a notable rate performance of 186.4 F/g at 20 A/g. Moreover, the specific-capacity retention values after 5,000 cycles stood at 93.8%, 89.8%, and 77.4% at 10, 15, and 20 A/g, respectively, as shown in Figure 6.15.

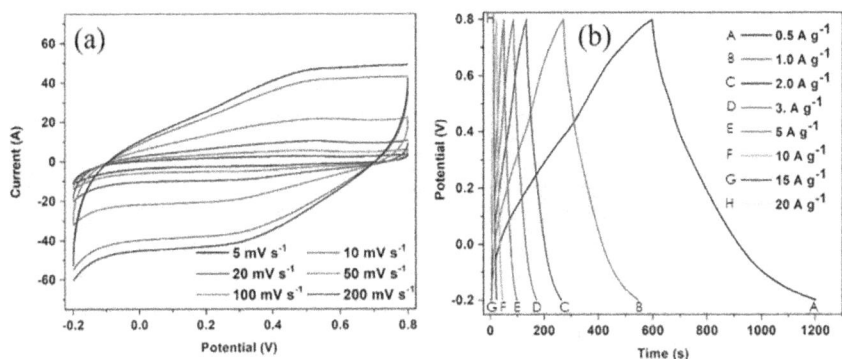

FIGURE 6.15 The electrochemical behavior of CQDs in supercapacitors: (a) through cyclic voltammetry (CV) curves illustrate the response of the CQD electrode across a range of scan rates (5–200 mV/s), covering a potential range of –0.2 to 0.8 V. These experiments were conducted in a 1 M H_2SO_4 electrolyte solution. (b) Galvanostatic charge-discharge (GCD) curves showcase the behavior of the CQDs electrode across various current densities ranging from 0.5 to 20 A/g. The experiments were conducted in a 1 M H_2SO_4 electrolyte solution. Adapted from Inayat et al. (2023).

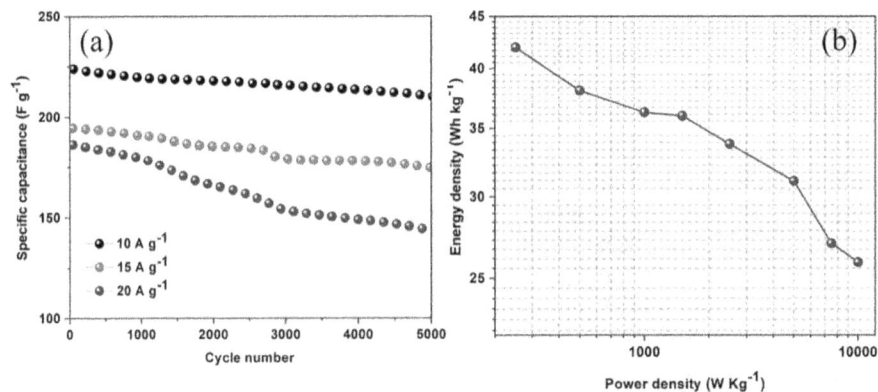

FIGURE 6.16 (a) The durability of the CQD electrode was evaluated through cyclic stability tests conducted over 5,000 cycles at current density levels of 10, 15, and 20 A/g. (b) The Ragone plot illustrates the performance of the CQD electrode across different current densities. Adapted from Inayat et al. (2023).

The notable preservation of high capacity can be attributed to the exceptional electronic and ionic conductivity of the CQDs, along with their structural adaptability, effectively preventing mechanical degradation during continuous insertion and extraction cycles. Additionally, the PD and ED reached impressive values of 25.8 and 10,000 Wh/kg, respectively, at a current of 20 A/g as shown in Figure 6.16. The introduction of nano-sized CQDs significantly improved the active sites of the particles, as well as surface roughness and electrical conductivity. This improvement further facilitated enhanced ion transfer and improved accessibility to electrolyte ions (Inayat et al., 2023).

The utilization of CQDs derived from spent tea leaves enables the integration of these CQDs into composite structures alongside transition-metal oxides and sulfides, thereby enhancing the properties of SCs. Exploring this avenue further holds the promise of generating a diverse range of valuable materials tailored for the advancement of SC technology.

In another study, H. Quan et al. conducted the synthesis of CQDs derived from Camellia Oleifera shell powder using hydrothermal method (Quan et al., 2022). Their findings indicated that CQDs integrated with hierarchical porous carbon (HPC) exhibited a notable specific surface area, well-suited pore structure, and favorable N- and O-containing functional groups. These attributes were found to be advantageous in enhancing active sites, as well as ameliorating the kinetics of charge transfer and the rate of ion transfer.

The resultant CQDs/HPC-2 material demonstrated remarkable electrochemical performance, with a substantial capacitance of 259 F/g at 1 A/g and exceptional rate capability in a 1 M H_2SO_4 electrolyte. Furthermore, when used in a symmetric SC configuration, CQDs/HPC-2 exhibited an impressive energy density of 8.61 Wh/kg at a power density of 477 W/kg in a 1 M H_2SO_4 solution, along with a remarkable energy density of 17.86 Wh/kg at a power density of 954 W/kg in a 1 M Na_2SO_4 solution as shown in Figure 6.17 (Quan et al., 2022).

The outcome of the research suggests that porous carbons derived from biomass hold significant promise in augmenting their energy storage capability, attributed to the uniform dispersion of highly crystallized GQDs. This approach introduces a novel perspective for fabricating electrode materials with remarkable performance, along with a novel strategy for the efficient utilization of biomass waste.

Chiu and Lin conducted a study where AC was produced from spent CGs through carbonization and activation methods using various activating agents (Chiu and Lin, 2019). They introduced a one-step activation process using six different agents, namely H_3PO_4, HCl, $FeCl_3$, $ZnCl_2$, NaOH, and KOH, to transform waste CGs into AC. They carefully analyzed and compared physical properties such as hydrophobic and hydrophilic functional groups, defect-to-graphene ratios, surface area, and pore volumes for AC produced with different activating agents.

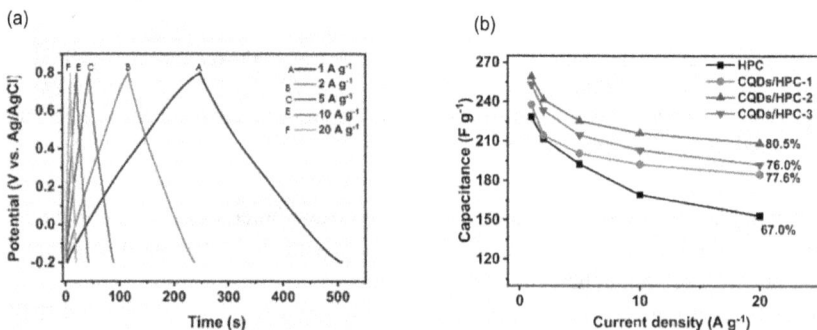

FIGURE 6.17   (a) Galvanostatic charge-discharge (GCD) curves of CQDs/HPC-2 were recorded under various current densities. (b) The specific capacitance of HPC electrodes was calculated across different current density values. Adapted from Quan et al. (2022).

The research found that the CV curve integrating area of the CG electrode was significantly smaller compared to the CV curves of AC electrodes. This indicated that the energy storage capability of the non-activated CG was notably inferior. The CV curves displayed increased oxidation and reduction currents at high and low potential regions, respectively, attributed to water-splitting reactions. Furthermore, GCD curves were presented for AC electrodes synthesized using various activating agents. These curves exhibited symmetric charge and discharge patterns with no plateaus, suggesting dominant EDLC behavior for energy storage in these AC electrodes. The AC electrodes prepared with different activating agents resulted in C_F of 51.0, 0.6, 41.3, 72.9, 69.5, and 105.3 F/g for H_3PO_4, HCl, $FeCl_3$, $ZnCl_2$, NaOH, and KOH, respectively. Notably, AC synthesized using KOH displayed the best performance with a C_F value of 105.3 F/g, surpassing the commercial carbon electrode with a C_F value of 21.8 F/g shown in Figure 6.18.

Zhang (2019) conducted an experiment in which they successfully synthesized mesoporous carbon materials, denoted as "Original-SE," through a high-temperature carbonization process using biomass materials derived from silkworms. These materials were developed for the purpose of serving as high-performance electrodes with a wide potential window in SC applications. They determined that under visible light exposure, they achieved a specific capacitance of 312 F/g at 0.1 A/g, surpassing more than twice the capacitance achieved in the absence of light (Zhang, 2019).

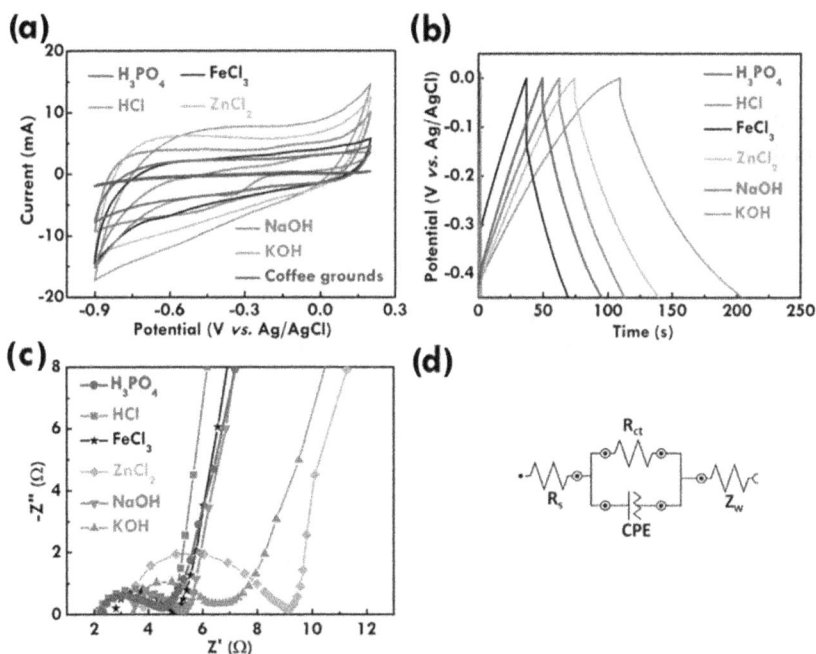

FIGURE 6.18 (a) The CV curves were obtained at a scan rate of 10 mV/s. (b) The GCD curves were recorded at a current density of 0.5 A/g. (c) The Nyquist plots were generated, and (d) the corresponding equivalent circuit was established for the AC electrodes synthesized using H_3PO_4, HCl, $FeCl_3$, $ZnCl_2$, NaOH, and KOH. Adapted from Chiu and Lin (2019).

TABLE 6.2
Electrochemical Behavior of CQDs or CDs in Supercapacitor

Electrode	Specific Capacitance	Energy Density	Power Density	References
Tea leaves	302 F/g	41.9 Wh/g	250 W/g	Inayat et al. (2023)
Spent coffee grounds	105.3 F/g	6.94 Wh/kg	350 W/Kg	Chiu and Lin (2019)
Silkworm	312 F/g	-	-	Zhang (2019)
Cauliflower	278 F/g	-	-	Hoang et al. (2019)
Zucchini	374 F/g	-	-	Hoang and Gomes (2019)

Meanwhile, Hoang et al. had outlined an eco-friendly, one-step synthesis method for producing CDs derived from biomass, using waste cauliflower leaves as a naturally occurring carbon source (Hoang et al., 2019). These synthesized CDs were then used to create composites with reduced graphene oxide (RGO) via a straightforward hydrothermal process. In this process, the CDs served as effective spacers, preventing the restacking of graphene nanosheets, which resulted in increased surface area and pore volume. The introduction of inherently nitrogen-doped CDs had a significant impact on various electrochemical properties. It led to reduced charge transfer resistance, improved proton diffusivity in acidic electrolytes, and the creation of active sites for pseudocapacitance due to the presence of oxygen and nitrogen functional groups. These alterations modified the electrochemical characteristics of RGO. The RGO/CD composite, with a GO:CD mass ratio of 2:1, demonstrated favorable specific capacitance and enhanced cycling stability. As the current density increased from 0.2 to 100 A/g and the scan rate rose from 2 to 1,000 mV/s, the highest discharge capacitance achieved was 278 F/g, with capacitance retention values of 60% and 63%, respectively (Hoang et al., 2019).

Separately, through an environmentally friendly and cost-effective process, Hoang and Gomes had successfully synthesized CDs from zucchini, utilizing waste zucchini biomass as a readily available and economical carbon precursor (Hoang and Gomes, 2019). The resulting CDs displayed remarkable electrochemical properties. At a scan rate of 2 m/Vs, a notable maximum specific capacitance of 374 F/g was achieved. The composite exhibited exceptional capacitance retention, reaching 71.7% as the scan rate was increased from 2 to 1,000 mV/s. Moreover, the composite demonstrated remarkable cycling stability, maintaining its capacitance over 10,000 cycles at 10 A/g with a retention rate of 93.8%. Table 6.2 presents an overview of the electrochemical characteristics of CQDs or CDs for SC, based on the data compiled from selected literature sources.

6.6 CONCLUSIONS AND PERSPECTIVES

The SCs and EDLCs, as well as a rundown of current developments, were presented in this article. According to the methodology they store charge, these SC designs can be split into three classes, including EDLC, pseudocapacitors, and hybrid capacitors. SCs have a lot of potential for applications due to their long storage life, strong cycling

TABLE 6.3
Comparing Energy Storing Technologies (Roy and Rengarajan, 2015; Arumugham et al., 2020; Sharma and Kumar, 2020)

Property	Battery	Supercapacitor
Energy density	2–100	1–10
Power density	50–200	7,000–18,000
Efficiency	85%–95%	>95%
Cycle life	10^5 times	>105 times

reliability, and short load time (Kiamahalleh et al., 2012). SCs must be made cheap while still keeping their extended cycle life and high concentration of energy. SCs feature up to 90% recycling stability and symmetric charge and discharge throughout their entire nominal voltage range, which is a significant benefit. As the SC has very low energy density compared to batteries (comparison shown in Table 6.3), one of the most important strategies for addressing the low energy density issue is the adoption of new, better materials for electrochemical SC electrodes (Conway, 1999).

We learned from the above discussion that numerous researchers have developed and synthesized CQDs using a variety of methodologies. Excellent characteristics of the CQDs include good biocompatibility, lower toxicity, fluorescent nature, straightforward synthesis, excellent electrical conductivity, and affordable price. CQDs have also been utilized in numerous significant applications, such as involving sensing, bio-medicine, optoelectronics, and the energy sector (Magesh et al., 2022). Based on the outcomes of this research, formulating innovative electrode materials using CQDs could serve as a viable and environmentally friendly approach for energy storage (Inayat et al., 2023).

In this review, we have outlined the applications of nanomaterials based on CQDs in SCs, with a focus on their fabrication, structure, and energy storage potential. Below is a synopsis of the key roles that CQDs fulfill in energy storage and conversion systems, leveraging their unique physical attributes in tandem with the current state of research.

1. Enhanced SC properties can be achieved through augmented specific surface area and nitrogen doping in conventional carbon materials, leading to improved rates and capacitance. This effect is attributed to the facilitation of electrolyte ion storage within deep pores.
2. Substantially reducing volume change stress and enhancing electrode dynamics can be accomplished by downsizing electrode materials to the scale of QDs.

Extensive efforts have been dedicated by researchers to formulate diverse synthesis methods for QDs. Furthermore, various design and assembly strategies have been established to enhance the quality of QD-based electrode materials. However, prior to transitioning to practical application, there remain numerous scientific and technical challenges within this field that require resolution.

Research on CQD composites in the field of electrochemical energy storage is still in its early stages. The lack of comprehensive studies on the electrochemical behavior and energy storage mechanisms of CQDs and other components hinders a complete understanding of how different CQD attributes impact their performance in energy storage and conversion systems. Given this scenario, conducting thorough investigations into aspects such as the interactions between CQDs and other materials, and their influence on composite material properties, is crucial. These investigations should encompass both theoretical and experimental analyses.

This research concludes by highlighting the significance and immense potential of quantum dot composites in the development of high-performance energy storage and catalytic systems. It can be deduced that CQDs are progressively emerging as essential multifunctional materials for energy storage and conversion technologies. With the advancement of state-of-the-art technologies and characterization methods, novel physicochemical attributes of CQDs will likely be uncovered, further broadening their utility across diverse research fields. We anticipate that the exploration of enhanced electrode materials based on CQDs and their potential utilization in energy-related industries will witness a rapid and extensive expansion in the near future.

ACKNOWLEDGMENT

The research was supported by the Ministry of Higher Education (MoHE), through the Fundamental Research Grant Scheme (FRGS/1/2022/TK08/UTAR/02/7).

REFERENCES

Abbas, A., Mariana, L. T., & Phan, A. N. (2018). Biomass-waste derived graphene quantum dots and their applications. *Carbon*, 140, 77–99.

Arora, N., & Sharma, N. N. (2014). Arc discharge synthesis of carbon nanotubes: Comprehensive review. *Diamond and Related Materials*, 50, 135–150.

Arumugham, T., Manikandan, A., Amimodu, R. G., Munusamy, S., & Iyer, S. K. (2020). A sustainable synthesis of green carbon quantum dot (CQD) from Catharanthus roseus (white flowering plant) leaves and investigation of its dual fluorescence responsive behavior in multi-ion detection and biological applications. *Sustainable Materials and Technologies*, 23, e00138.

Atchudan, R., Edison, T. N. J. I., Manivannan, S., Perumal, S., Somanathan, T., & Lee, Y. R. (2021). Sustainable synthesis of carbon quantum dots from banana peel waste using hydrothermal process for in vivo bioimaging. *Physical E-Low-Dimensional Systems & Nanostructures*, 126, 114417.

Azam, N., Ali, M. N., & Khan, T. J. (2021). Carbon quantum dots for biomedical applications: Review and analysis. *Frontiers in Materials*, 8: 700403.

Braga, R. F., Fernandes, D. M., Adán-Más, A., Silva, T. M., & Montemor, M. (2023). The role of the precursor on the electrochemical performance of N,S Co-Doped graphene electrodes in aqueous electrolytes. *Batteries*, 9(3), 168.

Chen, J., Du, H., Xu, Y., Ma, B., Zheng, Z., Li, P., & Jiang, Y. (2021). A turn-on fluorescent sensor based on coffee-ground carbon dots for the detection of sodium cyclamate. *Journal of Materials Science: Materials in Electronics*, 32(10), 13581–13587.

Chen, S., Ramachandran, R., Mani, V., & Saraswathi, R. (2014). Recent advancements in electrode materials for the highperformance electrochemical supercapacitors: A review. *International Journal of Electrochemical Science*, 9(8), 4072–4085.

Chiu, Y., & Lin, L. (2019). Effect of activating agents for producing activated carbon using a facile one-step synthesis with waste coffee grounds for symmetric supercapacitors. *Journal of the Taiwan Institute of Chemical Engineers*, 101, 177–185.

Choi, H., & Yoon, H. (2015). Nanostructured electrode materials for electrochemical capacitor applications. *Nanomaterials*, 5(2), 906–936.

Conway, B. E. (1999). *Electrochemical Supercapacitors*. Springer, New York.

Costa, A. I., Barata, P. D., De Moraes, B. G. S., & Prata, J. V. (2022). Carbon dots from coffee grounds: Synthesis, characterization and detection of noxious nitroanilines. *Chemosensors*, 10(3), 113.

Cui, L., Ren, X., Wang, J., & Sun, M. (2020). Synthesis of homogeneous carbon quantum dots by ultrafast dual-beam pulsed laser ablation for bioimaging. *Materials Today Nano*, 12, 100091.

De Klerk, N. J. J., Van Der Maas, E., & Wagemaker, M. (2018). Analysis of diffusion in solid-state electrolytes through md simulations, improvement of the li-ion conductivity in β-Li3PS4 as an example. *ACS Applied Energy Materials*, 1(7), 3230–3242.

Desmond, L. J., Phan, A. N., & Gentile, P. (2021). Critical overview on the green synthesis of carbon quantum dots and their application for cancer therapy. *Environmental Science. NANO*, 8(4), 848–862.

Dey, S., Govindaraj, A., Biswas, K., & Rao, C. N. R. (2014). Luminescence properties of boron and nitrogen doped graphene quantum dots prepared from arc-discharge generated doped graphene samples. *Chemical Physics Letters*, 595-596, 203–208.

Ezugbe, E. O., & Rathilal, S. (2020). Membrane technologies in wastewater treatment: A review. *Membranes*, 10(5), 89.

Feng, X., & Zhang, Y. (2019). A simple and green synthesis of carbon quantum dots from coke for white light-emitting devices. *RSC Advances*, 9(58), 33789–33793.

Glavin, M., & Hurley, W. (2012). Optimisation of a photovoltaic battery ultracapacitor hybrid energy storage system. *Solar Energy*, 86(10), 3009–3020.

Gonzalez, A., Goikolea, E., Barrena, J. A., & Mysyk, R. (2016). Review on supercapacitors: Technologies and materials. *Renewable & Sustainable Energy Reviews*, 58, 1189–1206.

Gualous, H., Chaoui, H., & Gallay, R. (2016). Supercapacitor calendar aging for telecommunication applications. In: *INTELEC, International Telecommunications Energy Conference (Proceedings)*. Institute of Electrical and Electronics Engineers Inc. Austin, TX, USA.

Gupta, S., Singh, P. K., & Bhattacharya, B. (2018). Low-viscosity ionic liquid-doped solid polymer electrolytes. *High Performance Polymers*, 30(8), 986–992.

Hashmi, S., Kumar, A., & Tripathi, S. K. (2007). Experimental studies on poly methyl methacrylate based gel polymer electrolytes for application in electrical double layer capacitors. *Journal of Physics D*, 40(21), 6527–6534.

He, M., Zhang, J., Wang, H., Kong, Y., Xiao, Y., & Xu, W. (2018). Material and optical properties of fluorescent carbon quantum dots fabricated from lemon juice via hydrothermal reaction. *Nanoscale Research Letters*, 13(1).

Hoang, V. C., & Gomes, V. G. (2019). High performance hybrid supercapacitor based on doped zucchini-derived carbon dots and graphene. *Materials Today Energy*, 12, 198- 207.

Hoang, V. C., Nguyen, L. H., & Gomes, V. G. (2019). High efficiency supercapacitor derived from biomass based carbon dots and reduced graphene oxide composite. *Journal of Electroanalytical Chemistry*, 832, 87–96.

Hu, L., Zhang, C., Zeng, G., Chen, G., Wan, J., Guo, Z., Wu, H., Yu, Z., Zhou, Y., & Liu, J. (2016). Metal-based quantum dots: Synthesis, surface modification, transport and fate in aquatic environments and toxicity to microorganisms. *RSC Advances* 6(82), 78595–78610.

Hu, S., Li, J., Yang, J., Wang, Y., & Cao, S. (2011). Laser synthesis and size tailor of carbon quantum dots. *Journal of Nanoparticle Research*, 13(12), 7247–7252.

Hu, S., Niu, K., Sun, J., Yang, J., Zhao, N., & Du, X. (2009). One-step synthesis of fluorescent carbon nanoparticles by laser irradiation. *Journal of Materials Chemistry*, 19(4), 484–488.

Inayat, A., Albalawi, K., Rehman, A. U., Adnan, A., Saad, A. Y., Saleh, E. a. M., Alamri, M. A., El-Zahhar, A. A., Haider, A., & Abbas, S. M. (2023). Tunable synthesis of carbon quantum dots from the biomass of spent tea leaves as supercapacitor electrode. *Materials Today Communications*, 34, 105479.

Iro, Z. S. (2016). A brief review on electrode materials for Supercapacitor. *International Journal of Electrochemical Science*, 10628–10643.

Jalal, N. I., Ibrahim, R. I., & Oudah, M. K. (2021). A review on supercapacitors: Types and components. *Journal of Physics*, 1973(1), 012015.

Jiang, K., Sun, S., Zhang, L., Lu, Y., Wu, A., Cai, C., & Lin, H. (2015). Red, green, and blue luminescence by carbon dots: Full-Color emission tuning and multicolor cellular imaging. *Angewandte Chemie*, 54(18), 5360–5363.

Kiamahalleh, M. V., Zein, S. H. S., Najafpour, G., Sata, S. A., & Buniran, S. (2012). Multiwalled carbon nanotubes based nanocomposites for supercapacitors: A review of electrode materials. *Nano*, 7(2), 1230002.

Kim, B. K., Sy, S., Yu, A., & Zhang, J. (2015). Electrochemical supercapacitors for energy storage and conversion. In: *Handbook of Clean Energy Systems*, Yan J. (ed.), 1–25.

Kraev, G., Hinov, N., Arnaudov, D., Rangelov, N. R., & Gilev, B. (2016). Serial ZVS DC-DC converter for supercapacitor charging. In: *2016 19th International Symposium on Electrical Apparatus and Technologies, SIELA 2016*. Institute of Electrical and Electronics Engineers Inc.

Kumari, R., Singh, V., & Ravikant, C. (2023). Enhanced performance of activated carbon-based supercapacitor derived from waste soybean oil with coffee ground additives. *Materials Chemistry and Physics*, 305, 127882.

Kuparowitz, T. Charge transport and storage in a supercapacitor structure. Brno University of Technology, The Faculty of Electrical Engineering and Communication. Ph.D. Thesis (2010)

Lawu, B. L., Fuada, S., Ramadhan, S., Sabana, A. F., & Sasongko, A. (2017). Charging supercapacitor mechanism based-on bidirectional DC-DC converter for electric ATV motor application. In: *2017 International Symposium on Electronics and Smart Devices, ISESD 2017*. Institute of Electrical and Electronics Engineers Inc, pp. 129–132.

Li, H., Kang, Z., Liu, Y., & Lee, S. (2012). Carbon nanodots: Synthesis, properties and applications. *Journal of Materials Chemistry*, 22(46), 24230.

Li, H., Tao, L., Huang, F., Sun, Q., Zhao, X., Han, J., Shen, Y., & Wang, M. (2017). Enhancing efficiency of perovskite solar cells via surface passivation with graphene oxide interlayer. *ACS Applied Materials & Interfaces*, 9(44), 38967–38976.

Li, X., Wang, H., Shimizu, Y., Pyatenko, A., Kawaguchi, K., & Koshizaki, N. (2011). Preparation of carbon quantum dots with tunable photoluminescence by rapid laser passivation in ordinary organic solvents. *Chemical Communications*, 47(3), 932–934.

Lim, S. Y., Shen, W., & Gao, Z. (2015). Carbon quantum dots and their applications. *Chemical Society Reviews*, 44(1), 362–381.

Liu, Y., Roy, S., Sarkar, S., Xu, J., Zhao, Y., & Zhang, J. (2021). A review of carbon dots and their composite materials for electrochemical energy technologies. *Carbon Energy*, 3(5), 795–826.

Magesh, V., Sundramoorthy, A. K., & Dhanraj, G. (2022). Recent advances on synthesis and potential applications of carbon quantum dots. *Frontiers in Materials*, 9: 906838.

Mahani, M., Pourrahmani-Sarbanani, M., Yoosefian, M., Divsar, F., Mousavi, S. M., & Nomani, A. (2021). Doxorubicin delivery to breast cancer cells with transferrin-targeted carbon quantum dots: An in vitro and in silico study. *Journal of Drug Delivery Science and Technology*, 62, 102342.

Monte-Filho, S. S., Andrade, S. I. E., Lima, M. B., & De Araújo, M. C. U. (2019). Synthesis of highly fluorescent carbon dots from lemon and onion juices for determination of riboflavin in multivitamin/mineral supplements. *Journal of Pharmaceutical Analysis*, 9(3), 209–216.

Najib, S., & Erdem, E. (2019). Current progress achieved in novel materials for supercapacitor electrodes: Mini review. *Nanoscale Advances*, 1(8), 2817–2827.

Prasad, G. G., Shetty, N., Thakur, S., Rakshitha, & Bommegowda, K. B. (2019). Supercapacitor technology and its applications: A review. *IOP Conference Series*, 561(1), 012105.

Quan, H., Tao, W., Wang, Y., & Chen, D. (2022). Enhanced supercapacitor performance of Camellia oleifera shell derived hierarchical porous carbon by carbon quantum dots. *Journal of Energy Storage*, 55, 105573.

Ren, W., Chen, S., Li, S., Zhang, Y., Liu, J., Guan, M., Yang, H., Li, N., Han, C., Li, T., Zhao, Z., & Ge, J. (2018). Photoluminescence enhancement of carbon dots by surfactants at room temperature. *Chemistry: A European Journal*, 24(59), 15806–15811.

Roy, B. P., & Rengarajan, N. (2015). Application of Supercapacitors for Short term Energy Storage Requirements. *Asian Journal of Applied Sciences*, 8(2), 158–164.

Sadat, A., Ding, H., Xu, M., Hu, X., Li, Z., Wang, J., Liu, L., Jiang, L., Wang, D., Dong, C., Yan, M., Wang, Q., & Hong, B. (2019). Recent advances in synthesis, optical properties, and biomedical applications of carbon dots. *ACS Applied Bio Materials*, 2(6), 2317–2338.

Şahin, M. E., & Blaabjerg, F. (2020). A hybrid PV-battery/supercapacitor system and a basic active power control proposal in MATLAB/Simulink. *Electronics*, 9(1), 129.

Sahin, M.E., Blaabjerg, F., & Sangwongwanich, A. (2020). A review on supercapacitor materials and developments. *Turkish Journal of Materials* 5, 10–24.

Salanne, M. (2017). Ionic liquids for supercapacitor applications. *Topics in Current Chemistry*, 375(3).

Sharma, P. R., & Kumar, V. (2020). Current technology of supercapacitors: A review. *Journal of Electronic Materials*, 49(6), 3520–3532.

Shen, P., & Xia, Y. (2014). Synthesis-modification integration: One-step fabrication of boronic acid functionalized carbon dots for fluorescent blood sugar sensing. *Analytical Chemistry*, 86(11), 5323–5329.

Shi, W., Zhu, J., Sim, D., Tay, Y. Y., Lu, Z., Zhang, X., Sharma, Y., Srinivasan, M., Zhang, H., Hng, H. H., & Yan, Q. (2011). Achieving high specific charge capacitances in Fe3O4/reduced graphene oxide nanocomposites. *Journal of Materials Chemistry*, 21(10), 3422.

Simon, P., & Gogotsi, Y. (2008). Materials for electrochemical capacitors. *Nature Materials*, 7(11), 845–854.

Sk, M. A., Ananthanarayanan, A., Huang, L., Lim, K. H., & Chen, P. (2014). Revealing the tunable photoluminescence properties of graphene quantum dots. *Journal of Materials Chemistry C*, 2(34), 6954–6960.

Thangaraj, B., Solomon, P. R., & Ranganathan, S. (2019). Synthesis of carbon quantum dots with special reference to biomass as a source: A Review. *Current Pharmaceutical Design*, 25(13), 1455–1476.

Tuerhong, M., Xu, Y., & Yin, X. (2017). Review on carbon dots and their applications. *Chinese Journal of Analytical Chemistry*, 45(1), 139–150.

Tungare, K., Bhori, M., Racherla, K. S., & Sawant, S. (2020). Synthesis, characterization and biocompatibility studies of carbon quantum dots from Phoenix dactylifera. *3 Biotech*, 10(12).

Wang, F., Wu, X., Yuan, X., Liu, Z., Zhang, Y., Fu, L., Zhu, Y., Zhou, Q., Wu, Y., & Huang, W. (2017). Latest advances in supercapacitors: From new electrode materials to novel device designs. *Chemical Society Reviews*, 46(22), 6816–6854.

Wang, X., Feng, Y., Dong, P., & Huang, J. (2019). A mini review on carbon quantum dots: Preparation, properties, and electrocatalytic application. *Frontiers in Chemistry*, 7: 671.

Wang, Y., & Hu, A. (2014). Carbon quantum dots: Synthesis, properties and applications. *Journal of Materials Chemistry C*, 2(34), 6921.

Wasterlain, S., Gualous, H., & Fauvarque, J.F. (2006). Hybrid power source with batteries and supercapacitor for vehicle applications. *Esscap2006*.

Xiao, W., Xia, H., Fuh, J., & Li, L. (2009). Growth of single-crystal α-MnO2 nanotubes prepared by a hydrothermal route and their electrochemical properties. *Journal of Power Sources*, 193(2), 935–938.

Yang, Z., Li, Z., Xu, M., Ma, Y., Zhang, J., Su, Y., Gao, F., Wei, H., & Zhang, L. (2013). Controllable synthesis of fluorescent carbon dots and their detection application as nanoprobes. *Nano-Micro Letters* 5, 247–259.

Yuan, F., Ding, L., Li, Y., Li, X., Fan, L., Zhou, S., Fang, D., & Yang, S. (2015). Multicolor fluorescent graphene quantum dots colorimetrically responsive to all-pH and a wide temperature range. *Nanoscale*, 7(27), 11727–11733.

Zhai, X., Zhang, P., Liu, C., Bai, T., Li, W., Dai, L., & Liu, W. (2012). Highly luminescent carbon nanodots by microwave-assisted pyrolysis. *Chemical Communications*, 48(64), 7955.

Zhang, C., Zhu, F., Xu, H., Liu, W., Yang, L., Ma, J., Kang, Z., & Liu, Y. (2017). Significant improvement of near-UV electroluminescence from ZnO quantum dot LEDs via coupling with carbon nanodot surface plasmons. *Nanoscale*, 9(38), 14592–14601.

Zhang, Q., Sun, X., Hong, R., Yin, K., & Li, H. (2017). Production of yellow-emitting carbon quantum dots from fullerene carbon soot. *Science China. Materials*, 60(2), 141–150.

Zhang, X. (2019). Mesoporous biochar derived from silkworm excrement for high performance supercapacitors. *International Journal of Electrochemical Science*, 14(9), 8793–8804.

Zuo, P., Xiuhua, L., Sun, Z., Guo, Y., & He, H. (2015). A review on syntheses, properties, characterization and bioanalytical applications of fluorescent carbon dots. *Mikrochimica Acta*, 183(2), 519–542.

7 Tin-Based Anodes for Next-Generation Lithium-Ion Batteries

L. P. Teo, M. H. Buraidah, and A. K. Arof

7.1 INTRODUCTION

Batteries are inseparable from modern daily lives. They are found in small items such as watches, kitchen scales, clocks, remote controls, toy cars, calculators, mobile phones, and laptops, and in larger objects including cars, electric buses, and satellites, as well as in space applications.

Batteries, in general, are distributed into two broad classifications: primary and secondary. The former is non-rechargeable and must be disposed of after a single usage, whereas the latter are rechargeable and can be used multiple times. A very good representative of secondary batteries is lithium-ion batteries (LIBs). Apart from the above-mentioned items, LIB can also be found in iPads, digital cameras, video camcorders, hybrid and electric vehicles, and so on.

Basically, LIBs are electrochemical cells that produce electrical energy from their stored chemical energy. The main constituents of LIBs are the anode and cathode, along with the electrolyte and separator separating the two electrodes. Here, we focus solely on the anode. The anode of a LIB is the component that donates electrons to the exterior circuit and is oxidized when the electrochemical reaction occurs. An ideal anode in an LIB should have features such as the ability to intercalate/deintercalate lithium ions at low potential with great capacity, diffuse Li$^+$ ions quickly and conduct them fast, undergo minimal structural and redox potential variation during intercalation/deintercalation of Li$^+$ ions (Tao et al., 2011). Anodes should also form a good solid-electrolyte interface (SEI) with the electrolyte during the initial cycle and remain compatible in subsequent cycles.

Essentially, anode materials can be sectionalized into two categories: carbon and non-carbonaceous materials. Graphite can be acknowledged as the most commonly used carbon-based anode owing to its distinctive features, which include (1) abundance in nature, (2) environmental benignity, (3) easy intercalation/deintercalation of Li$^+$ ions into its structure, (4) flat and low working potential versus lithium, (5) good cycle life, and (6) relatively cheap (Zhang, 2011; Tao et al., 2011; Manthiram, 2011; Chen, 2013). The theoretical specific capacity of an anode can be obtained from the following formula:

$$\text{Theoretical capacity} = \frac{\text{Faraday constant (Ah/mol)}}{3,600\,\text{s}} \div \left(\text{Molar mass} \times \text{Lithiated number}\right)$$

As an example, let's consider graphite. Graphite has a molar mass of 12 g/mol and is a type of carbon material consisting of 6 C atoms, which can lithiate with 1 Li. Based on this, it is known that its theoretical capacity is 372 mAh/g. Therefore, it is crucial to identify how many atoms of the active material will be lithiated with 1 Li. To ease readers' understanding, the following Table 7.1 provides the values of the lithiated phase of some anode materials along with their properties.

Graphite also has other disadvantages, such as (1) low energy density, (2) a low Li^+ ion insertion potential (0.05 V versus lithium), resulting in lithium adhering to the surface of graphite and potentially leading to safety issues, (3) a low Li^+ ion chemical diffusion coefficient (10^{-7} to $10^{-9} cm^2/s$), and (4) the possibility of organic solvent co-intercalation in graphite (Wakihara, 2001; Zhang, 2011; Tao et al., 2011; Manthiram, 2011; Zhang et al., 2013; Goriparti et al., 2014). These disadvantages have motivated researchers to explore alternative anode materials, such as silicon (Si), aluminum (Al), tin (Sn), magnesium (Mg), lithium titanate ($Li_4Ti_5O_{12}$), and various transition metal oxides like NiO, Co_3O_4, Fe_2O_3, etc. In this chapter, special attention is given to tin-based compounds, which show promising potential as anodes for the next generation of LIB. Herein, some achievements with tin-based materials used as anodes in LIB are presented.

From the periodic table, tin is positioned in Group IV along with other elements: carbon (C), silicon (Si), germanium (Ge), and lead (Pb). Although silicon possesses a higher theoretical capacity than tin (Table 7.1), the latter is less brittle and has higher electrical conductivity compared to the former (Zhao et al., 2015). In addition, tin-based materials are cheaper, abundant in nature, environmentally acceptable, and easier to process compared to silicon-based materials (Kamali and Fray, 2011; Chen, 2013; Wan et al., 2014).

On the other hand, germanium is expensive, whereas lead is toxic, making it less attractive. Moreover, tin is non-toxic, and its working potential of approximately 0.6 V versus Li makes it safe and attractive without compromising the operating voltage of the cell. In 1997, Idota and et al. (1997) pronounced that amorphous tin composite oxide containing metallic elements, i.e., boron (B), phosphorus (P), and aluminum (Al), attained greater capacity (1,030 mAh/g) than graphite. This amorphous tin-based oxide delivered ~100% coulombic efficiency with a reversible capacity of 650 mAh/g upon the initial charge in the potential window 0–1.2 V versus Li/Li^+. It has also been reported that the amorphous nanostructured Sn–Co–C anode

TABLE 7.1

Suitable Anode Materials in Lithium-Ion Batteries (Wu and Cui, 2012)

Materials	Li	C	Si	Sn	Sb	Al	Mg
Density (g cm^{-3})	0.53	2.25	2.23	7.29	6.7	2.70	1.30
Lithiated phase	Li	LiC_6	$Li_{4.4}Si$	$Li_{4.4}Sn$	Li_3Sb	LiAl	Li_3Mg
Theoretical capacity (mAh/g)	3,862	372	4,200	994	660	993	3,350
Volume change (%)	100	12	320	260	200	96	100
Potential vs. Li (V)	0	0.05	0.4	0.6	0.9	0.3	0.1

exhibited a 30% increment in volumetric capacity compared to a typical carbon anode (Whittingham, 2008). This makes tin-based materials more attractive for study. The Sn–Co–C anode has been marketed by Sony.

One crucial benefit of tin over graphite and carbon-based materials is that there is no co-intercalation of solvents in tin-containing substances, making it safe (Hu et al., 2012). Besides this, tin also has a higher theoretical capacity and relatively high lithium insertion potential compared to graphite (Wang et al., 2011; Yoon and Manthiram, 2011).

Tin is known to belong to a tetragonal crystal structure with lattice constants, a = 0.5831 nm and c = 0.3182 nm (Hu et al., 2012). As mentioned earlier, the theoretical capacity of tin without any impurity is 994 mAh/g, which is about three times greater than graphite (372 mAh/g), based on the final lithiated phase $Li_{4.4}Sn$. Tin can interact with lithium to form numerous lithium alloys, namely $LiSn$, Li_2Sn_5, Li_3Sn_2, Li_5Sn_2, Li_7Sn_2, Li_7Sn_3, and $Li_{22}Sn_5$ following the reaction (Winter and Besenhard, 1999; Zhang et al., 2019):

$$Li_xSn \leftrightarrow xLi^+ + xe^- + Sı$$

Crystallographic studies have suggested that $Li_{17}Sn_4$ (4.25 Li per Sn) is the realistic form of the end lithiated phase and not the previously assumed $Li_{22}Sn_5$ phase (Chen, 2013). Hence, its maximum gravimetric capacity is 959.5 mAh/g, which is still much higher than that of graphite. It has been reported that pure tin foil (bulk) delivered a reversible capacity of 600 mAh/g for only 10–15 cycles (Yang et al., 2003). After 15 cycles, the expansion and contraction of the electrode crystalline lattice occurred, resulting in an increase in cell impedance and a decrement in capacity. An electroplated Sn thin film on copper foil exhibited an initial capacity of 860 mAh/g and a reversible capacity of 200 mAh/g after 20 cycles (Morimoto and Tobishima, 2005). The highest capacity of 940 mAh/g has been recorded for electroplated tin, which is close to the theoretical capacity (Morimoto and Tobishima, 2005).

However, a pure tin anode undergoes a large volume change (260%) due to lithium-ion insertion/extraction (Kamali and Fray, 2011). This causes breakup/cracking of the tin anode, limiting its cycle life and resulting in an increase in cell impedance. To overcome this problem, tin in nanostructured form, tin-based alloys, and nanocomposites containing tin have been extensively studied, as a smaller particle size can help prevent the aggregation of tin into large clusters. Generally, when the size of tin particles is at the nanometer level, the lithium-ion diffusion length can be significantly reduced, thus reducing the volume expansion/contraction during repeated cycling while maintaining structural integrity.

7.2 TIN-BASED ALLOYS

By dispersing tin particles in an inactive matrix, the volume change problem can be alleviated. The electrochemically conducting and mechanically ductile inactive matrix provides a structurally stable form to counter the volume change of tin and decrease irreversible capacity loss. Many intermetallic alloys, such as Cu_6Sn_5, $FeSn_2$, $FeSn$, and others, have shown improved performance compared to pure tin.

$CoSn_3$, $FeSn_2$, Ni_3Sn_4, and Cu_6Sn_5 have theoretical capacities of 852, 804, 725, and 605 mAh/g, respectively, to name a few. Kepler et al. (1999) reported that Cu_6Sn_4 exhibited better performance compared to Cu_6Sn_5 and Cu_6Sn_6 for all 20 cycles under a constant current of 0.1 mA in the potential window 0–1 V. On the first cycle, the Cu_6Sn_4 cell delivered a capacity of ~450 mAh/g and ~150 mAh/g at the 20th cycle, while the Cu_6Sn_6 cell exhibited capacities of ~350 mAh/g and ~80 mAh/g on the first and 20th cycles, respectively. The Cu_6Sn_5 cell delivered a capacity of ~345 mAh/g at the end of the first cycle and ~70 mAh/g at the end of the 20th cycle. Zou and Wang (2013) prepared tin sulfide (SnS_2) via a microwave-assisted solvothermal technique, which delivered capacities of 1,755 and 370 mAh/g at the first and 50th cycles, respectively, at a current density of 100 mA/g in the voltage range from 0.005 to 1.1 V. A recent review in the literature is dedicated to tin sulfides (Mou et al., 2020). Zhang et al. (2008) reported that both $FeSn_2$ prepared by solvothermal and chemical reduction methods exhibited initial discharge capacities of ~640 mAh/g and ~750 mAh/g at a current density of 80 mA/g. However, solvothermal-prepared $FeSn_2$ maintained a reversible discharge capacity of ~435 mAh/g, whereas $FeSn_2$ prepared via the chemical reduction technique yielded a capacity of ~400 mAh/g at the 20th cycle. In another report, an LIB using $FeSn_2$ nanospheres attained a capacity of ~500 mAh/g for almost 15 cycles continuously (Wang et al., 2012a). The commercialization of the Nexelion battery, which consists of amorphous nano-SnCo inserted in carbon, delivered an initial capacity of 300 mAh/g but then decreased to ~200 mAh/g after 10 cycles (Fan et al., 2007). Sn/Fe/C composite made from mechanical milling utilizing Al, Ti, and Mg as reducing agents with various grinding media exhibited a specific capacity of 600 mAh/g (Zhang and Whittingham, 2010). The Sn-Fe carbon cell also showed good capacity retention upon cycling for 200 cycles, reaching a specific capacity of 250 mAh/g in the potential window between 0.01 and 1.5 V at a current density of 5 mA/cm^2. It is interesting to note that Dang and co-workers (2015) prepared micro-$Sn_{0.9}Se_{0.1}$ alloy via the jet-milling method instead of nanosize, maintaining a capacity of ~500 mAh/g after 100 cycles at a 0.5 C rate.

7.2.1 TIN-BASED OXIDES

Tin-based oxide materials have garnered considerable attention due to their stability, ease of synthesis with varying morphologies, and controllable particle sizes. These oxides are known for exhibiting reduced capacity fading compared to pure tin. In this context, we will explore various tin oxide materials used as anodes in LIBs.

7.2.2 TIN (II) OXIDE (SNO)

SnO crystallizes in a tetragonal structure with lattice parameters, $a = b = 3.803$ Å and $c = 4.849$ Å (Macías et al., 2011). From the literature, it can be understood that the reaction mechanism between SnO and Li$^+$ ion can be summarized as follows (Uchiyama et al., 2008):

- $SnO + 2Li^+ + 2e^- \rightarrow Sn + Li_2O$
- $Sn + xLi^+ + xe^- \leftrightarrow Li_xSn \ (0 \leq x \leq 4.4)$

As seen in the first equation, the electrochemical reaction of SnO with Li$^+$ ions leads to the formation of metallic tin and Li$_2$O, which is an irreversible process. After that, the metallic tin reacts reversibly with Li$^+$ ions to form Li$_x$Sn alloy, as shown in the second equation. SnO is said to possess a high theoretical specific capacity of 875 mAh/g (Uchiyama et al., 2008). However, the synthesis of SnO is challenging since Sn^{2+} can easily oxidize to Sn^{4+}, making it less studied compared to its counterpart, SnO$_2$.

Zhang and co-workers succeeded in obtaining ultrathin SnO nanosheets using a facile one-step hydrothermal method, utilizing tin dichloride dihydrate (SnCl$_2$·2H$_2$O) and hexamethylenetetramine as precursors (Zhang et al., 2014). The SnO nanosheets delivered high specific discharge capacities of 1,496, 882, 559, and 305.4 mAh/g at the first, fifth, 20th, and 40th cycles, respectively, under a constant specific current of 100 mA/g in the potential window between 0.005 and 3 V. Even at higher current densities of 200 and 500 mA/g, the SnO nanosheets achieved capacities of 785.4 and 606.3 mA/g, respectively, indicating their good rate capability (Zhang et al., 2014). The SnO nanosheets have a short diffusion length for Li$^+$ ion insertion and a large surface area to facilitate fast Li$^+$ ion transport. In addition, the space between the nanosheets, built up in a disordered manner, can act as a buffer against volume expansion/contraction during Li+ ion insertion/de-insertion (Zhang et al., 2014).

SnO prepared in the form of nanoplatelets was reported for the first time using a reducing atmosphere flame spray pyrolysis technique without a catalyst and surfactant (Hu et al., 2014). The SnO nanoplatelets, which are in a rounded square shape and have an average diameter of ~250 nm with a thickness of ~30 nm, exhibited initial discharge and charge capacities of ~1,900 mAh/g and ~950 mAh/g, respectively, under a specific current drain of 100 mA/g between 0.005 and 2 V. The discharge capacity was maintained at 330 mAh/g after 40 cycles. Similarly, hydrothermal-derived SnO nanoflowers exhibited an initial discharge capacity of ~2,000 mAh/g and charge capacity of 950 mAh/g in the potential window between 0.01 and 2 V, with a reversible capacity of ~200 mAh/g at the end of the 25th cycle (Iqbal et al., 2012). Therefore, it can be inferred that the morphology of SnO will affect the electrochemical performance. It has been reported that SnO morphology can be altered by using different solvents in the wet chemistry route (Jaśkaniec et al., 2021). SnO prepared using solvents, viz., ethanol, deionized water, 1-hexanol, and a mixture of water/methanol (30:70 ratio), displayed flower-like, square-shaped, cracked squares, and platelets-like morphologies (Jaśkaniec et al., 2021). Their capacity performance decreases in the order of SnO (methanol) > SnO (ethanol) > SnO (hexanol) > SnO (deionized water) at a 0.1C rate.

Although there is relatively little research involving SnO compared to research on SnO$_2$, SnO shows good electrochemical performance in some cases. Thomas and Rao (2014) reported that SnO nanoplates embedded in graphene nanosheets (GNSs) exhibited better performance than SnO$_2$ nanoparticles surrounded by GNS. The two tin oxide-based materials were deposited by the evaporation of tin granules in oxygen. The phase and morphology of the tin oxide-based materials were controlled by the evaporation rate of tin granules. The initial discharge capacity of 4,004 and 2,461 mAh/g were obtained by SnO-GNS and SnO$_2$-GNS, respectively. SnO-GNS also showed a higher reversible capacity of 1,022 mAh/g than that of SnO$_2$-GNS

(715 mAh/g) after 40 cycles at a current density of 23 μA/cm². The unique characteristics of GNS, such as high flexibility, good conductivity, a large surface, and chemical stability, make it a good matrix to embed the tin oxide materials to reduce strain during lithiation/de-lithiation processes and thus minimize volume change.

On the other hand, Hong and Kang (2015) prepared core-shell-structured tin oxide-carbon composite powders via a one-pot spray pyrolysis using a solution containing tin (II) oxalate (C_2O_4Sn) and polyvinylpyrrolidone (PVP). The tin oxide-carbon (SnO_x-C) composite with a mole ratio $SnO:SnO_2$ of 24:76 (6:19) exhibited an initial discharge capacity of 1,667 mAh/g, whereas bare tin dioxide (SnO_2) only delivered 1,473 mAh/g at a current density of 1 A/g. At a higher current density of 2 A/g, the tin oxide-carbon composite showed a stable discharge capacity of 1,033 mAh/g at the end of the 500th cycle, but the discharge capacity of bare SnO_2 decreased continuously to 78 mAh/g at the 500th cycle. SnO_x-C also showed better rate capability than bare SnO_2 when cycled at different current densities. The excellent performance of the tin oxide-carbon composite can be attributed to the low porosity property of carbon, which enables the formation of a stable SEI layer and thus prevents the electrolyte from penetrating into the composite (Hong and Kang,, 2015).

7.2.3   Tin (IV) Oxide (SnO_2)

Similar to SnO, SnO_2 also has a tetragonal crystal structure and lattice parameters of $a = b = 4.738$ Å and $c = 3.188$ Å (Chandra Bose et al., 2002; Liu et al., 2017). SnO_2 forms an alloy with lithium in a Li/SnO_2 half-cell in which the reactions can be written as follows (Sivashanmugam et al., 2005; Kim et al., 2014):

- $SnO_2 + 4Li^+ + 4e^- \rightarrow Sn + 2Li_2O$
- $Sn + xLi^+ + xe^- \leftrightarrow Li_xSn \ (0 < x < 4.4)$

The first equation illustrates the irreversible reduction of tin dioxide to tin, which is dispersed in a lithium oxide (Li_2O) matrix during the first discharge process. The Li_2O matrix is said to help alleviate volume change but not for prolonged cycling (Wang et al., 2011). On the other hand, the second equation indicates the reversible alloying/dealloying processes, where the metallic tin reacts with Li-ion to produce $Li_{4.4}Sn$ alloy. This shows that one Sn atom is capable of storing 4.4 Li, resulting in a theoretical specific capacity of 783 mAh/g. Experimentally, an initial capacity of ~1,689 mAh/g was obtained by SnO_2 nanoparticles prepared via the sol-gel technique at a constant current of 0.1 C in the 0.01–2.00 V range (Liu et al., 2017). The capacity declined to ~571 mAh/g after the 20th cycle. Ultrafine SnO_2 nanoparticles recorded a discharge capacity of ~1,197 mAh/g and ~273 mAh/g at the first and 50th cycles, respectively, under a 100 mA/g current density (Yin et al., 2016). Using the pulse laser deposition technique, SnO_2 thin film anode first attained the specific capacity of ca. 642 mAh/g and then decreased progressively from 400 mAh/g (100th cycle) to 200 mAh/g (250th cycle) before maintaining to this value for the next 6,250 cycles under a 2 C current density in the potential window 0.01–3 V (Dai et al., 2021). Nonetheless, SnO_2 is still affected by volume change. Therefore, the performance of SnO_2 can be improved by coating with carbon (Wang et al., 2011) and polypyrrole

(PPy) as well as doping with copper (Cu), nitrogen (N), indium (In), to name a few (Wan et al., 2014;).

Deposition of a small quantity of cobalt (Co) element into SnO_2 can hinder particle aggregation and thus enhances lithium-storage capability (Mei et al., 2012). Co-SnO_2 delivered high capacities of 1,956, 1,346, and 810 mAh/g at the first, second, and 50th cycles, which are higher than that of pure SnO_2 (first cycle—1,584 mAh/g, second cycle—983 mAh/g, 50th cycle—499 mAh/g). Nanocrystalline $Sn_{0.9}In_{0.1}O_2$ showed better performance with a reversible capacity of ~500 mAh/g at the 10th cycle than the undoped nano-SnO_2 (~300 mAh/g) (Subramanian et al., 2004). Wan et al. (2014) revealed that co-doping of Cu and N into SnO_2 via the hydrothermal method exhibited enhanced capacity retention when compared to Cu-doped SnO_2 and pure SnO_2. The charge capacities of SnO_2, Cu/SnO_2, and Cu-N/SnO_2 were recorded at 1,624, 1,846, and 1,939 mAh/g, respectively, at the first cycle and 200, 462, and 664 mAh/g after the 50th cycle. It is worth mentioning that antimony-doped SnO_2 nanopowders can maintain a reversible capacity of 637 mAh/g at the end of the 100th cycle with a high initial discharge capacity of 2,400 mAh/g (Wang et al., 2009a). Fluorine-doped SnO_2 nanoparticles attached to reduced graphene oxide (RGO) sheets have been reported with good electrochemical properties (100 mA/g: 1,037 mAh/g at the 150th cycle; 500 mA/g: 733 mAh/g at the 250th cycle; 1 A/g: 860 mAh/g; 2 A/g: 770 mAh/g) (Cui et al., 2017). In another report, SnO_2 nanoparticles have been doped with molybdenum (Mo) and then implanted in graphite nanosheets by Feng et al. (2020). This design produced astonishing results (1,317 mAh/g capacity at the 200th cycle under a 0.2 A/g current density; 759 mAh/g capacity at the 950th cycle under a 1 A/g current density).

Carbon coated on SnO_2 acts as a shield, minimizing large volume alteration and improving electronic conductivity, resulting in better performance (Wang et al., 2011). Carbon-coated SnO_2 hollow microspheres, produced via the hydrothermal method, showed improved cycling performance than uncoated SnO_2 (Guo et al., 2013). At a current density of 100 mA/g and a potential window between 0.005 and 3 V, the initial discharge and charge capacities were ~1,650 mAh/g and ~900 mAh/g for SnO_2 hollow microspheres. Under similar conditions, SnO_2 coated with carbon exhibited an irreversible discharge capacity of ~1,580 mAh/g and a charge capacity of ~1,100 mAh/g in the first cycle. After cycling 20 times, the carbon-coated SnO_2 delivered a discharge capacity of ~680 mAh/g, which was ~180 mAh/g more than uncoated SnO_2 at the same cycle. The hollow structure of SnO_2 has provided prereserved space to accommodate the volume expansion (Guo et al., 2013). A three-dimensional (3D) nanostructured SnO_2-carbon composite encapsulated in copper foam attained a slow capacity decline from a reversible capacity of ~1,099 mAh/g (second cycle) to 846 mAh/g (100th cycle) at a 0.2 A/g current density in the voltage window 0.01–1.50 V (Chen et al., 2016). Wang et al. (2021) confined nanoparticles of SnO_2 and Sn in a porous carbon network and recorded high-capacity retention of 1,105 mAh/g at the 290th cycle under a current rate of 200 mA/g. The electrode also showed decent performance at different current rates (500 mA/g: 453 mAh/g; 5,000 mA/g: 171 mAh/g; 1 A/g: 940 mAh/g; 10 A/g: 107 mAh/g) (Wang et al., 2021). In another paper, a ternary design of Sn with SnO_2 in the carbon matrix presented an initial reversible capacity of 2,248 mAh/g (100 mA/g) and a 475 mAh/g capacity at the 100th cycle (500 mA/g) (Saddique et al., 2022).

On the other hand, conducting polymers, namely PPy and polyaniline (PANI), can also help alleviate the volume expansion by providing a soft matrix that can accommodate the stress induced during lithium insertion and extraction (Cui et al., 2011). However, conducting polymers have only moderate conductivity, whereas the conductivity of SnO_2 is low (Huang et al., 2014). Therefore, Huang et al. (2014) prepared hierarchical SnO_2 with double carbon coating, which showed higher capacity and better cycling performance compared to uncoated SnO_2 and single coating of SnO_2 with PPy-derived carbon and RGO. The SnO_2 was first coated with carbon derived from PPy and then with RGO. Under the same specific current drain of 250 mA/g and potential window between 0.005 and 3 V, uncoated SnO_2 only exhibited a capacity of 37.4 mAh/g, whereas the capacities of 161.6, 291, and 550 mAh/g were obtained for PPy-derived carbon/SnO_2, RGO/SnO_2, and RGO/PPy-derived carbon/SnO_2, respectively, at the end of the 50th cycle (Huang et al., 2014). Recently, Shin and co-workers (2022) prepared an anode from porous nano-SnO_2 coated with PANI on a Cu substrate that can deliver a capacity of 1,644, 1,002, and 440 mAh/g at the first, second, and 50th cycles under a current density of 0.1 A/g and voltage region 0.01–3 V. No binder was used in such an electrode by the authors. The nano-SnO_2 anode without PANI can only achieve ~200 mAh/g at the 50th cycle, which is attributed to its higher charge-transfer resistance than the PANI-coated SnO_2 electrode (Shin et al., 2022).

In addition to coating and doping, researchers have also explored different morphologies of SnO_2, such as nanorods, nanowires, nanotubes, nanosheets, or hollow or porous nanostructures as ways to relieve the severe volume change problem. Nanostructured SnO_2 in the forms of wire, rod, tube, and sheet can withstand the large volume variation during Li+ ion insertion/extraction processes. Nano-SnO_2 in a one-dimensional (1D) structure (nanowires, nanorods, nanotubes) shows a large surface area to volume ratio, good electrical and optical characteristics, better mechanical stability, efficient electron transport, and excellent capacity.

7.2.4 SnO_2 NANOWIRES

Park et al. (2008) examined the morphology effect on the electrochemical performance of SnO_2 nanowires, SnO_2 nanotubes, and SnO_2 nanoparticles. Under similar conditions, i.e., a voltage window from 0.05 to 1.50 V and a specific current of 100 mA/g, both SnO_2 nanowires and nanotubes exhibited a higher reversible specific capacity of 300 and 100 mAh/g, respectively, compared to SnO_2 nanopowders with an average particle size of 100 nm at the end of the 15th cycle. SnO_2 nanowires, prepared by the self-catalysis growth technique with diameters between 200 and 500 nm, demonstrated improved performance compared to SnO_2 in powder form, attributed to the high surface area that provided more reaction sites in the nanowires (Park et al., 2007). Following the report by Meduri et al. (2009), SnO_2 nanowires achieved an initial capacity of 2,400 mAh/g with a reversible capacity of 166 mAh/g at the end of the 40th cycle. In comparison, a capacity of 2,800 and 490 mAh/g at the first cycle and 40th cycle, respectively, was observed for metal tin nanoclusters embedded in SnO_2 nanowire networks. Using the atomic layer deposition technique, Guan and co-workers (2014) prepared nanowired SnO_2 surrounded by TiO_2 nanotubes, with carbon cloth as the substrate. The TiO_2 nanotubes aimed to stabilize the SEI layer and endure the volume expansion. This formation achieved better electrochemical performance in terms of rate capability and cyclability compared to the electrode with SnO_2 nanowires only. SnO_2 nanowires doped with antimony (Sb) and coated with a nanocarbon shield exhibited excellent capacity retention, with a 585 mAh/g capacity at the 500th cycle under 100 mA/g, compared to anodes containing Sb-doped SnO_2 (152 mAh/g) and SnO_2 nanowires only (52 mAh/g) (Mousavi et al., 2021).

7.2.5 SnO_2 Nanorods

Similar to nanowires, the nanorod morphology is advantageous for the good cycling performance of SnO_2 due to its 1D structure, which can provide high inter-porosity and sufficient space to sustain the volume change. It is worth mentioning that Wen and co-workers (Wen et al., 2013) employed the hydrothermal technique to produce flower-like SnO_2 nanorod packets formed by adding oleic acid to a mixture of sodium hydroxide (NaOH) and tin tetrachloride pentahydrate ($SnCl_4 \cdot 5H_2O$). The flower-like SnO_2 nanorod bundles exhibited a first discharge capacity of 1,673 mAh/g and a charge capacity of 815 mAh/g in the potential window from 0 to 2.5 V at a 0.1C rate (78 mA/g). Upon continuous cycling for 40 cycles, its capacity could still maintain 694 mAh/g, which is ~2 times higher than that of graphite (Wen et al., 2013). In another report, the initial discharge capacity of SnO_2 nanorods by the hydrothermal method almost reached 1,600 mAh/g at a 0.2C rate in the potential window 0.01–3.00 V (Zhang et al., 2019). Nonetheless, a capacity of ~400 mAh/g was obtained after the 40th cycle. There is a controversial report that revealed SnO_2 nanorods performed poorly (641 mAh/g) than SnO_2 nanoparticles (526 mAh/g) upon continuous cycling for 50 times at a 200 mA/g current density (Hameed et al., 2017). The former also suffered from a bigger irreversible capacity loss (first discharge capacity 1,529 mAh/g; first charge capacity 872 mAh/g) than the latter (first discharge capacity 1,529 mAh/g; first charge capacity 872 mAh/g), although it showed a higher initial discharge capacity. This is attributed to the lower value of the lithium diffusion coefficient ($4.36 \times 10^{-6} cm^2/s$) obtained by SnO_2 nanorods as compared to SnO_2 nanoparticles ($9.14 \times 10^{-6} cm^2/s$) (Hameed et al., 2017).

Nanorods of SnO_2 have been adhered to carbon to avoid direct contact with the electrolyte to prevent SEI cracks and showed better electrochemical performance (initial capacity 1,881 mAh/g, 884 mAh/g at 10th cycle) compared to SnO_2 nanorods without carbon at 100 mA/g current density and a voltage range of 0.005–1.5 V (Palaniyandy et al., 2019). Yu and co-workers (2013) have reported that hydrothermal-derived SnO_2 nanorods delivered discharge and charge capacities of 1,963 and 862 mAh/g at a 100 mA/g current density with a voltage range of 0.005–3 V. The charge and discharge capacities of 167 and 626 mAh/g were lower for carbon-coated SnO_2 nanorods under similar cycling conditions. After 10 times cycling at a specific current of 100 mA/g, the uncoated and carbon-coated SnO_2 exhibited reversible capacities of 690 and 653 mAh/g, respectively. Thereafter, the uncoated and carbon-coated SnO_2 were cycled at a specific current of 500 mA/g for another 10 cycles. It was found that the capacity of uncoated SnO_2 decreased from 474 mAh/g at the 11th cycle to 374 mAh/g at the 20th cycle, whereas carbon-coated SnO_2 showed stable cycling performance, maintaining the capacity between 190 and 200 mAh/g throughout the cycling. It is interesting to note that carbon coating in this case does not show improved performance as expected but instead shows capacity decline and poor rate capability. The authors have indicated that the lower conductivity of carbon-coated SnO_2 nanorods (4.2×10^{-6} S/m) than that of uncoated SnO_2 (9.9×10^{-4} S/m) may be the reason behind this (Yu et al., 2013). Another style of SnO_2 in the form of nanoribbons has been synthesized on a stainless-steel foil and used directly as an anode without a binder (Faramarzi et al., 2018). The initial irreversible capacity of ~1,818 mAh/g was obtained, followed by a sharp decline to ~929 mAh/g, thereafter gradually decreased to ~676 mAh/g after 20 times cycling at a 300 mA/g current density in the potential range 0.05–2.00 V. The authors have claimed that nano-SnO_2 in a ribbon arrangement showed less capacity fading than SnO_2 in nanowire form due to smaller thickness, shorter Li-ion diffusion length, and a large surface area (Faramarzi et al., 2018).

7.2.6 SnO_2 Nanotubes

The nanotube morphology has a beneficial effect on the electrochemical performance of SnO_2, where its 1D tubular structure restricts the movement of SnO_2, and thus prevents the SnO_2 particles from agglomerating. Also, the porous surface and cavities of the nanotubes provide empty space that can accommodate the structural strain and withstand the volume change. Ye et al. (2010) investigated the size effect of SnO_2 nanotubes on electrochemical properties. They found that smaller SnO_2 nanotube sizes prepared via the hydrothermal method using mesostructured silica nanorods as a template give high reversible capacity and good cycling performance (468 mAh/g after 30 cycles), which may be due to its more robust structure. In contrast, long SnO_2 nanotubes can only retain a capacity of ~395 mAh/g at the end of the 30th cycle. Zhu et al. (2015) designed an approach for preparing carbon-coated SnO_2 nanotubes by a hydrothermal route using zinc oxide (ZnO) nanorods as a template. The process involves forming ZnO-SnO_2 nanorods from the template ZnO, followed by hydrothermal carbonization with glucose as the carbon precursor to obtain carbon-coated SnO_2 nanotubes. The carbon-coated SnO_2 nanotubes have a capacity of ~1,711 mAh/g, ~834 mAh/g, ~454 mAh/g, and ~406 mAh/g at the first, second, 50th, and 100th cycles, respectively.

7.2.7 SnO_2 Nanosheets

SnO_2 nanosheets have a two-dimensional (2D) structure that offers structural flexibility to accommodate the large volume change upon lithium insertion/extraction. However, it is quite complex to prepare well-defined SnO_2 nanosheets due to the difficulty in controlling the reaction rate of the oxidation process of Sn^{2+} to Sn^{4+}, anisotropic crystal growth, and the hydrolysis of tin salt (Chen and Lou, 2012). Liu et al. (2013) prepared SnO_2 nanosheets by a hydrothermal route without utilizing both surface-active agents and organic solvents. The material showed good electrochemical performance with a large reversible capacity and excellent cycling stability. Cycling between 0 and 3 V, the specific capacity of SnO$_2$ nanosheets was maintained at 604 mAh/g over 50 cycles at a specific current of 100 mA/g, with initial discharge and charge capacities of 1,508 and 1,053 mAh/g, respectively. Flower-like spheres assembled from SnO$_2$ nanosheets have been reported by Wu et al. (2011) to possess good rate capability and a high reversible capacity of 400 mAh/g under a high current density of 800 mA/g. The authors believed that good electrochemical performance was attributed to a shorter lithium diffusion path. Other SnO$_2$ nanosheets also displayed promising performance at different current rates (731 mAh/g at 1 C; 678 mAh/g at 2 C; 424 mAh/g at 4 C) (Wang et al., 2021). It is worth mentioning that the SnO$_2$ nanosheets were very thin (ca. 20 nm) and utilized straightaway as an electrode without mixing with carbon black super P and polyvinylidene fluoride (PVDF) binder. Also, it is noteworthy that SnO$_2$ nanopillars prepared via pulsed laser deposition showed a 3D irregular cylindrical structure with a pyramid shape at the upper part, and accomplished admirable long-term cycling performance (Dai et al., 2021). SnO$_2$ nanopillars deposited on copper foil directly without PVDF binder exhibited 1,082, 832, 524, and 313 mAh/g at first, 300th, 1,100th, and 6,500th cycle, respectively, under 2 C in the 0.01–3 V range. Furthermore, good rate cyclability is also shown by this material (Dai et al., 2021). In short, such a design allows more space for expansion.

Wang et al. (2019) reported that SnO$_2$ nanospheres-implanted RGO forming a 2D architecture was capable of delivering the initial capacities of 1,212 and 1,063 mAh/g under 0.1 and 1 A/g specific currents, respectively, in the 0.01–3 V window. This anode material achieved a reversible capacity of 1,335 mAh/g at 1 A/g specific current after 500 times cycling, thus demonstrating superior long-term cycling at high current density. Another anode material showing good long-cycle life with high-rate capability was made from SnO$_2$ hollow nanospheres coated with PPy (Yuan et al., 2017). At various current rates of 100, 200, 500, and 1,000 mA/g, the mean specific capacities were 368, 307, 229, and 169 mAh/g, respectively. After cycling at different current densities, the electrode managed to obtain a capacity of 408 mAh/g when subjected again to a specific current of 100 mA/g. Under the same rate, ~900 mAh/g capacity was attained after 600 times cycling (Yuan et al., 2017). As expected, the nano-SnO$_2$ electrode without PPy coating performed poorly. The conducting polymer offers mechanical integrity and good conductivity. A more detailed review on SnO$_2$ (Mou et al., 2020) and specifically on SnO$_2$ nanosheets (Chen and Lou, 2012) is available elsewhere.

7.2.8 SnO$_2$/Carbon Composites

Whether it is coating or incorporating carbon into SnO$_2$, carbon provides good electronic conductivity to the SnO$_2$ and also buffers against volume expansion during Li$^+$ insertion. Hence, this will improve the capacity and cyclability. There are many ways to prepare SnO$_2$/carbon composite. Liu et al. (2012) synthesized SnO$_2$ nanorods grown on graphite by a hydrothermal method, which showed better electrochemical performance. The SnO$_2$/graphite nanocomposites have an initial discharge capacity of 1,197 mAh/g and a charge capacity of 872 mAh/g at a specific current of 50 mA/g and a voltage window between 0 and 3 V. However, under similar conditions, the lithium insertion capacity of SnO_2 nanorods is higher (~1,650 mAh/g) than that of both SnO_2/graphite and graphite (~600 mAh/g). Nevertheless, after 10 cycles, SnO_2/graphite delivers the highest reversible capacity of 627 mAh/g compared to graphite and SnO_2, with specific capacities of ~390 and ~300 mAh/g, respectively. The SnO_2/graphite composite also shows a high specific capacity at higher rates with a capacity of ~600 mAh/g at 100 mA/g and ~410 mAh/g at 500 mA/g, much higher than graphite and SnO_2. The graphite in the composite can obstruct SnO_2 nanorods from agglomeration and thus enhances the lithium-storage properties.

Chen and Yano (2013) designed and prepared a nanocomposite containing 80 wt.% SnO_2 nanoparticles and 20 wt.% carbon spheres via a simple infiltration procedure. An irreversible discharge capacity and charge capacity of 2,038 and 1,468 mAh/g were obtained at the first cycle, and the capacity was retained at 1,200 mAh/g after 30 cycles at a specific current of 80 mA/g and a potential window between 0.005 and 3 V. The high surface area mesoporous carbon spheres have open pore channels that enable the electrolyte to penetrate into SnO_2 and give fast Li^+ ion diffusion. The empty space at the interior of carbon can tolerate the volume expansion. Using the microwave-hydrothermal technique, Liu et al. (2015) prepared SnO_2/graphene composite which exhibits first discharge and charge capacities of 2,213 and 1,402 mAh/g at a specific current of 100 mA/g with a cutoff voltage window between 0.01 and 3 V. At the end of the 100th cycle, capacity remained at 1,359 mAh/g. At a large specific current of 1,000 mA/g, the SnO_2/graphene composite can still deliver a high discharge capacity of 1,502 and 677 mAh/g at the first and 1,000th cycles, respectively.

Ternary nanocomposite containing SnO_2-titanium dioxide (TiO_2) in GNSs has been prepared by a simple reflux method. It exhibits better performance with an initial discharge capacity of ~1,595 mAh/g and a reversible capacity of ~1,500 mAh/g at 100 mA/g between 0.01 and 3 V (Chen and Qin, 2014). A specific capacity of ~1,177 mAh/g can still be delivered even after 100 cycles. On the contrary, the discharge capacity of the binary SnO_2-graphene nanocomposite is ~1,384 mAh/g and ~208 mAh/g at the first and 100th cycles, respectively. The incorporation of TiO_2 hinders the cracking and prevents the loss of interparticle contact and indirectly improves the stability of SnO_2 during volume change, whereas the function of GNSs is as mentioned above. A much more comprehensive review of SnO_2-based nanomaterials can be found in references (Goriparti et al., 2014; Zhao et al., 2015; Zoller et al., 2019).

7.3   TERNARY TIN OXIDES

To the best of our knowledge, there is no single article that summarizes the literature review on almost all ternary tin oxides and their electrochemical properties, unlike SnO_2. Nonetheless, one recent article focuses only on three tin-based materials: manganese orthostannate (Mn_2SnO_4), cobalt orthostannate (Co_2SnO_4), and zinc orthostannate (Zn_2SnO_4), along with their applications in LIB and sodium-ion batteries (Duan et al., 2023). Another article provides an overview specifically for Zn_2SnO_4

and zinc metastannate ($ZnSnO_3$) (Sun and Liang, 2017). The latter emphasizes the synthesis–morphology relationship with a brief description of LIB. Herein, we provide a brief account of ternary tin-based oxide materials, namely lithium tin oxide (Li_2SnO_3), calcium metastannate ($CaSnO_3$), barium metastannate ($BaSnO_3$), cobalt metastannate ($CoSnO_3$), strontium metastannate ($SrSnO_3$), cadmium metastannate ($CdSnO_3$), calcium orthostannate (Ca_2SnO_4), magnesium orthostannate (Mg_2SnO_4), Mn_2SnO_4, Co_2SnO_4, and Zn_2SnO_4, as anodes in Li-ion batteries.

7.3.1 LITHIUM TIN OXIDE OR LITHIUM STANNATE (Li_2SnO_3)

According to the literature (Lang, 1966; Tarakina et al., 2009), lithium stannate, or lithium tin oxide (Li_2SnO_3), exists in two forms: α-Li_2SnO_3, stable up to 800°C, and β-Li_2SnO_3, produced at 1,000°C. β-Li_2SnO_3 crystallizes in a monoclinic structure (space group of C2/c) with lattice parameters $a = 5.289\,Å$, $b = 9.187\,Å$, $c = 10.026\,Å$, $\beta = 100.35°$ (Hodeau et al., 1982; Fu et al., 2016). Table 7.2 summarizes the electrochemical performance of Li_2SnO_3 reported by many researchers. It is evident from the table that the synthesis method affects the electrochemical performance. Li_2SnO_3 prepared by the sol-gel technique exhibited better performance than that prepared via solid-state reaction. This difference may be attributed to the morphology of the Li_2SnO_3 product. Li_2SnO_3 derived from the sol-gel method has uniform nanocrystallites with a narrow particle size distribution (200–300 nm), whereas an agglomerate of particles with different sizes is observed for Li_2SnO_3 prepared by the solid-state reaction method. In the sol-gel method, a homogeneous mixing of the starting materials can occur at the atomic or molecular level, resulting in a decrease in the product's particle size (Lin et al., 2013). On the other hand, in the solid-state reaction method, the reactants are simply ground together in an agate mortar and undergo a heating process. As seen in Table 7.2, it is noteworthy that the hydrothermal-derived Li_2SnO_3 exhibited the highest specific capacity at the first cycle and still attained a capacity of 613 mAh/g after 30 cycles (Wang et al., 2013), compared to those prepared via sol-gel and solid-state reaction techniques.

To further improve the electrochemical properties of Li_2SnO_3, researchers have explored coating and doping Li_2SnO_3 with materials such as PPy (Huang et al., 2012), carbon (Wang et al., 2012b, 2012c), graphene (Zhao et al., 2013a), and PANI (Wang et al., 2012d). A composite of bamboo charcoal/Li_2SnO_3, prepared via the sol-gel technique, exhibited an improvement in capacity (1,689 mAh/g at the first cycle and ~674 mAh/g at the 30th cycle) compared to a Li//Li_2SnO_3 cell with a capacity of ~1,196 and 202 mAh/g at the first and 30th cycles, respectively, under a specific current of 60 mA/g in the potential window from 0.01 to 2 V (Zhao et al., 2013b). The porous structure and large surface area of bamboo charcoal can reduce volume expansion/contraction during Li^+ intercalation/deintercalation reactions. Wang et al. (2012b) first synthesized Li_2SnO_3 by the sol-gel method before doping it with carbon via a carbothermic reduction process. The carbon-doped Li_2SnO_3 exhibited an initial discharge capacity of 1,671 mAh/g and maintained a reversible capacity of 576.9 mAh/g at the end of the 50th cycle. It is worth highlighting that carbon-doped Li_2SnO_3 has been employed by Wang et al. (2012b) as a cathode material instead of an anode material for LIBs.

TABLE 7.2

Literature Review on Li$_2$SnO$_3$

Method	Starting Materials	Capacity (mAh/g)	Potential Window (V)	Current or Current Density (mA or mA/g)	References
Solid-state reaction	SnO$_2$ and Li$_2$CO$_3$	1,200-first cycle; 550-second cycle	0–2.5	9.3 mA/g	Courtney and Dahn (1997)
Solid-state reaction	SnO$_2$ and Li$_2$CO$_3$ Calcined at 1,000°C	1,177-first cycle	0.01–2	0.05 mA	Belliard and Irvine (2001)
Solid-state reaction and ball-milling	SnO$_2$ and Li$_2$CO$_3$ Calcined at 650°C	1,167-first cycle	0.01–2	0.05 mA	Belliard and Irvine (2001)
Solid-state reaction	SnO$_2$ and Li$_2$CO$_3$ – ball-milled 15 h – first calcination at 800°C for 6 hours – second calcination at 1,000°C for 12 h	1,074-first cycle; ~363-30th cycle	0–1	20 mA/g-first cycle; 60 mA/g subsequent cycle	Zhang et al. (2006)
Sol-gel	SnCl$_4$·5H$_2$O and Li$_2$CO$_3$ – citric acid as complexing agent	1,157-first cycle; ~400-30th cycle	0–1	20 mA/g-first cycle; 60 mA/g subsequent cycle	Zhang et al. (2006)
Solid-state reaction	SnC$_2$O$_4$ and Li$_2$CO$_3$ – heating 900°C 12 h	1,250-first cycle; ~400-10th cycle	0–1.5	0.1 mA	Vaughey et al. (2007)
Hydrothermal	SnCl$_4$·5H$_2$O and LiOH	2,105-first cycle; ~613-30th cycle	0–2	60 mA/g	Wang et al. (2013)
Hydrothermal	SnCl$_4$·5H$_2$O and LiOH.H$_2$O – calcined at 800°C for 6 hours	850-first cycle; <50-200th cycle	0.01–3	150 mA/g	Wang et al. (2019)
Hydrothermal	SnO$_2$ and LiOH.H$_2$O – vacuum drying at 60°C overnight	2,582-first cycle; ~179-50th cycle	0.1–2	0.1C	Zakuan et al. (2019)

As reported by Zhao et al. (2013a), Li_2SnO_3 powders distributed on GNSs demonstrated good cycling performance with a reversible capacity of 582 mAh/g at the 50th cycle at 60 mA/g, which is higher than the capacity of GNSs (408 mAh/g). It is said that the GNSs are flexible and can accommodate the volume change, thereby improving the structural stability of the composite. On the other hand, a slight increment in Li-ion storage capability with a reversible capacity of 560 mAh/g after 50 cycles has been observed in PPy-coated Li_2SnO_3 compared to that of uncoated Li_2SnO_3 (510 mAh/g) (Huang et al., 2012). A similar observation in Li_2SnO_3/PANI nanocomposites (569 mAh/g at the 50th cycle) and pure Li_2SnO_3 (510 mAh/g at the 50th cycle) has been reported (Wang et al., 2012a). Both PPy and PANI, which are conducting polymers, can serve as a backbone for Li_2SnO_3, while their soft structure acts as a shield to Li_2SnO_3 in relieving the volume change during discharge/charge processes. According to literature (Yang et al., 2019), nanocomposites comprised of Li_2SnO_3, SnO_2, and Li_2SiO_3 produced improved cycling and rate performance than Li_2SnO_3 only due to the former having a better framework to alleviate volume expansion and improve electrode–electrolyte contact. Another work has reported an anode made of Li_2SnO_3, Li_2SiO_3, and graphene which gave ~1,017 mAh/g capacity at a current density of 150 mA/g in the voltage window 0.01–3 V (Wang et al., 2019). Lower capacity values were obtained for Li_2SnO_3-Li_2SiO_3 (935 mAh/g) and Li_2SnO_3 (860 mAh/g) electrodes. Li_2SnO_3-Li_2SiO_3-graphene electrode also showed slow capacity loss after 200 times cycling compared to the other two. It should be noted that in the work of Wang et al. (2019), both Li_2SnO_3 and Li_2SiO_3 were the active materials and contributed to the electrochemical process. A full cell has also been fabricated using the Li_2SnO_3-Li_2SiO_3-graphene anode and $LiMn_2O_4$ cathode. At a current rate of 40 mA/g, the recorded capacities were 263 and ~92 mAh/g at the first and 50th cycles, respectively (Wang et al., 2019).

In order to further enhance the electrochemical performance, a ternary structure incorporating carbon-based materials and conducting polymers in addition to Li_2SnO_3 has been proposed. Li-ion half cells employing ternary Li_2SnO_3 composites, i.e., carbon-doped Li_2SnO_3/graphene and graphene/Li_2SnO_3/PPy, have been reported to exhibit better electrochemical properties compared to binary composites and pure Li_2SnO_3 material (Zhao et al. 2013c, 2013d). Reversible capacities of ~825 and 699 mAh/g have been obtained for carbon-doped Li_2SnO_3/graphene and graphene/ Li_2SnO_3/PPy at the end of the 30th cycle under a current density of 60 mA/g in the potential window 0–2 V (Zhao et al., 2013c, 2013d). The improved performance of ternary Li_2SnO_3 composite is attributed to the presence of carbon/graphene and PPy, which can serve as a double protection for Li_2SnO_3 when volume change takes place during Li^+ ion insertion/extraction. It is interesting to note that nano-Li_2SnO_3 has been coated on $LiNi_{0.5}Co_{0.2}Mn_{0.3}O_2$ cathode and displayed better performance than the non-coated $LiNi_{0.5}Co_{0.2}Mn_{0.3}O_2$ cathode (Hu et al., 2016). According to the authors, Li_2SnO_3 worked as a shield to safeguard $LiNi_{0.5}Co_{0.2}Mn_{0.3}O_2$ particles from electrolyte decomposition. Other than that, it has been reported that Li_2SnO_3 prepared via vapor-solid technique showed luminescent properties and thus has potential in light-emitting devices (García-Tecedor et al., 2019).

7.3.2 CALCIUM METASTANNATE (CaSnO₃)

The crystal structure of $CaSnO_3$ is orthorhombic, with a unit cell having lattice parameters $a = 11.09$ Å, $b = 7.91$ Å, and $c = 11.34$ Å (Sharma et al., 2002). Table 7.3 lists the research conducted by other scientists on LIBs using $CaSnO_3$. It can be observed from Table 7.3 that a higher capacity was obtained in $CaSnO_3$ prepared by the sol-gel technique compared to that prepared by the solid-state reaction method. This is due to the uniform particle size of sol-gel-derived $CaSnO_3$, whereas solid-state-prepared $CaSnO_3$ has large clusters of particles with different sizes. The uniformity of the particle size results in a smaller diffusion length, thus leading to a higher capacity. On the other hand, $CaSnO_3$ prepared via the hydrothermal technique, which showed a porous flower-like morphology (Zhao et al., 2010), exhibited a higher capacity at the 30th cycle under a specific current drain of 60 mA/g compared to sol-gel-derived and solid-state-prepared $CaSnO_3$ at the same cycle (Sharma et al., 2002). It is interesting to note that by using different concentrations of PVP as a surfactant, different morphologies of $CaSnO_3$ can be produced, although the starting materials used in the hydrothermal method are the same (Kim et al., 2014). Higher capacity is found in cubic-shaped $CaSnO_3$, which was prepared using 5 mL PVP, than the star-shaped $CaSnO_3$ produced with 1 mL PVP. Based on these observations, it can be concluded that the morphology of the materials influences the electrochemical properties. Carbon-coated $CaSnO_3$ nanotubes have been prepared by the solvothermal method to achieve better electrochemical properties (Hu et al., 2012). The carbon-coated $CaSnO_3$ nanotubes exhibited an initial capacity of 1,296 mAh/g at a specific current of 60 mA/g, whereas the uncoated $CaSnO_3$ nanotubes gave only a capacity of ~1,020 mAh/g. At the 30th cycle, the uncoated $CaSnO_3$ nanotubes delivered a capacity of 60 mAh/g, whereas ~190 mAh/g was achieved by the carbon-coated $CaSnO_3$ in the potential range of 0.005–1.3 V. It is said that coating with amorphous carbon acts as a barrier to relieve the volume expansion-contraction and also helps to maintain the dimensional stability of the electrode during the alloying/dealloying reaction (Hu et al., 2012).

7.3.3 STRONTIUM METASTANNATE (SrSnO₃)

Table 7.4 summarizes the methods of preparation for $SrSnO_3$ and the electrochemical properties in LIB applications available in the literature. It is worth mentioning that $SrSnO_3$ in powder form, prepared by both sol-gel and solid-state reaction methods, crystallizes in an orthorhombic crystal structure with lattice parameters $a = 5.71$ Å, $b = 5.69$ Å, and $c = 8.05$ Å (Sharma et al., 2005). However, Li et al. (2011) synthesized $SrSnO_3$ nanorods via cetyltrimethylammonium bromide (CTAB)-assisted hydrothermal method, which has a cubic structure with a lattice constant, $a = 8.04$ Å. The $SrSnO_3$ nanorods that formed by subjecting the precursor, i.e., $SrSn(OH)_6$ nanowires, to hydrothermal treatment in ethanol at 180°C for 8 hours and calcined in air at 700°C for 3 hours display the orthorhombic crystal structure (Hu et al., 2010). The $SrSn(OH)_6$ nanowires were previously prepared via the sonochemical method at room temperature without a surfactant. The $SrSnO_3$ nanorods prepared by this method exhibit an impurity peak of strontium carbonate ($SrCO_3$), but it does not

TABLE 7.3
Preparation Methods of CaSnO₃ and Their Electrochemical Performance in Li-Ion Batteries

Method	Starting Materials	Morphology	Capacity (mAh/g)	Potential Window (V)	Current or Current Density (mA or mA/g)	References
Solid-state reaction	SnO_2 and $Ca(NO_3)_2 \cdot 4H_2O$	Large clusters of different sizes particles	~1,054-first cycle; ~265–30th cycle	0.005–1	10 mA/g-first cycle; 60 mA/g subsequent cycle	Sharma et al. (2002)
Sol-gel	$CaCO_3$ and $SnCl_2$	Uniform nanocrystallites (200–300 nm)	~1,083-first cycle; ~340–30th cycle	0.005–1	10 mA/g-first cycle; 60 mA/g subsequent cycle	Sharma et al. (2002)
Hydrothermal	$Ca(NO_3)_2 \cdot 4H_2O$ and $Na_2SnO_3;3H_2O$	Porous flower-like	1,725-first cycle; 544–30th cycle	0.005–1	60 mA/g	Zhao et al. (2010)
Hydrothermal	$CaCl_2$ and Na_2SnO_3 – 1 mL PVP as surfactant	Star-shaped particles with size of ~μm	~872-first cycle; ~228–30th cycle	0–1	50 mA/g	Kim et al. (2014)
Hydrothermal	$CaCl_2$ and Na_2SnO_3 – 5 mL PVP as surfactant	Cubic-shaped with smaller particles size (~500 nm)	~1,238-first cycle; ~386–30th cycle	0–1	50 mA/g	Kim et al. (2014)
Electrospinning	$SnCl_2 \cdot 2H_2O$ and $Ca(NO_3)_2 \cdot 4H_2O$ Metal: PAN = 1.30:1	Eggroll-like nanotubes Average diameter: 560 nm	~1,153-first cycle; ~648–50th cycle	0.005- 3	60 mA/g	Li et al. (2013)
Electrospinning	$SnCl_2 \cdot 2H_2O$ and $Ca(NO_3)_2 \cdot 4H_2O$ Metal: PAN = 0.65:1	Nanorods average diameter: 380 nm	~1,177-first cycle; ~410–50th cycle	0.005-3	60 mA/g	Li et al. (2013)

$Ca(NO_3)_2 \cdot 4H_2O$, calcium nitrate tetrahydrate; $CaCO_3$, calcium carbonate; $SnCl_2$, tin (II) chloride; $Na_2SnO_3 \cdot 3H_2O$, sodium stannate trihydrate; $CaCl_2$, calcium chloride; Na_2SnO_3, sodium stannate; PVP, polyvinylpyrrolidone; $SnCl_2 \cdot 2H_2O$, tin (II) chloride dihydrate; PAN, polyacrylonitrile.

TABLE 7.4

Preparation Methods of SrSnO$_3$ and Their Electrochemical Performance in Li-Ion Batteries

Method	Starting Materials	Morphology	Capacity (mAh/g)	Potential Window (V)	Current Density (mA/g)	References
Solid-state reaction	SnO$_2$ and SrCO$_3$	Well-separated smooth particles with average size of 0.5–1 μm	738-first cycle; 144-20th cycle; 82-40th cycle	0.005–1	10 mA/g-first cycle; 30 mA/g-20th cycle; 60 mA/g-40th cycle	Sharma et al. (2005)
Sol-gel	SrCO$_3$ and SnCl$_2$	Aggregates of particles with size 200–300 nm	870-first cycle; 222-20th cycle; 184-40th cycle	0.005–1	10 mA/g-first cycle; 30 mA/g-20th cycle; 60 mA/g-40th cycle	Sharma et al. (2005)
Hydrothermal	SnCl$_4$·2H$_2$O and Sr(NO$_3$)$_2$ – CTAB as surfactant	Nanorods with diameter 100–150 nm	1,400-first cycle; ~150-10th cycle	0–2	80 mA/g	Li et al. (2011)
Hydrothermal	SrSn(OH)$_6$ nanowires	Nanorods with uniform diameter of ~100 nm	714-first cycle; ~200-50th cycle	0.005–2	50 mA/g	Hu et al. (2010)

SrCO$_3$, strontium carbonate; SnCl$_2$·2H$_2$O, tin (IV) chloride dihydrate; Sr(NO$_3$)$_2$, strontium nitrate; CTAB, cethyltrimethyl ammonium bromide; SrSn(OH)$_6$, strontium tin hydroxide.

involve in the alloying/de-alloying reactions. It is also interesting to point out that in ref (Li et al., 2011), $SrSnO_3$ in the nanorods form can be produced at hydrothermal temperatures of 160°C and above with CTAB as a surfactant. At a temperature of 140°C, $SrSnO_3$ exists in the micelles form (Li et al., 2011). Therefore, in this case, it can be inferred that the hydrothermal reaction temperature and CTAB plays an important role in influencing the crystal structure formation of $SrSnO_3$. CTAB is known to act as a template for the 1D directional growth of nanostructured material (Li et al., 2011). On the other hand, Hu et al. (2010) have proven that $SrSnO_3$ nanorods achieve better electrochemical performance than $SrSnO_3$ in the bulk form since the nanorods can tolerate the stress that occurs during volume changes during charging/ discharging processes.

7.3.4   BARIUM METASTANNATE ($BASNO_3$)

Barium metastannate ($BaSnO_3$) exhibits a cubic structure with a lattice constant a of 4.11 Å. It has found applications in gas sensors, ceramic capacitors, dielectric ceramics, and photocatalytic processes, in addition to its use in LIB (Nithyadharseni et al., 2016). Table 7.5 lists the synthesis processes of $BaSnO_3$, the starting materials used, and their electrochemical properties as reported in the literature. It is noteworthy that $BaSnO_3$, when prepared by the solid-state reaction method, exhibited a higher capacity than that synthesized via the sol-gel technique, regardless of the current density. This result contrasts with the $CaSnO_3$, $SrSnO_3$, and Li_2SnO_3 systems, where sol-gel-derived materials demonstrated better electrochemical performance than those prepared by the solid-state reaction method, as observed in Tables 7.2–7.5. It is also observed in Table 7.5 that $BaSnO_3$ exhibits fast capacity fading when cycling at 0.005–3.0 V compared to $BaSnO_3$ in the potential window of 0.005–1.0 V. However, a higher capacity of $BaSnO_3$ can be seen at the end of the 50th cycle in the potential range of 0.005–3.0 V (Nithyadharseni et al., 2016). As is customary for performance improvement, $BaSnO_3$ was mixed with RGO (Veerappan et al., 2016). The nanocomposites of $BaSnO_3$ and RGO showed the best performance in terms of the highest capacity, rate capability, and cycling stability when compared to pure $BaSnO_3$ and RGO.

7.3.5   COBALT METASTANNATE ($COSNO_3$)

Huang et al. (2003) synthesized $CoSnO_3$ using a pyrolysis reaction with starting materials, i.e., cobalt (II), sulfate ($CoSO_4$), and sodium stannate (Na_2SnO_3). They found that $CoSnO_3$ is amorphous in nature and has a square flake shape with an average particle size of 70 nm. Li//$CoSnO_3$ cells delivered an initial irreversible capacity of ~1,840 mAh/g and a reversible capacity of ~1,050 mAh/g at the 10th cycle under a specific current of 100 mA/g in a 0–3 V potential window (Huang et al., 2003). Additionally, the authors observed that $CoSnO_3$ prepared by the solid-state route suffered from a fast capacity degradation compared to pyrolysis-made $CoSnO_3$ (Huang et al., 2003). In another report, $CoSnO_3$ synthesized from a solvothermal method gave an initial capacity of ~1,408 mAh/g at a 50 mA/g current rate in a 0.005–3 V voltage window (Cao et al., 2014). When the current density varied to

TABLE 7.5

Preparation Methods of BaSnO$_3$ and Their Electrochemical Performance in Li-Ion Batteries

Method	Starting Materials	Morphology	Capacity (mAh/g)	Potential Window (V)	Current Density (mA/g)	Reference
Solid-state reaction	SnO$_2$ and BaCO$_3$	–	721-first cycle; 190-20th cycle; 142-40th cycle	0.005–1	10 mA/g-first cycle; 30 mA/g-20th cycle; 60 mA/g-40th cycle	Sharma et al. (2005)
Sol-gel	BaCO$_3$ and SnCl$_2$	–	715-first cycle; 156-20th cycle; 122-40th cycle	0.005–1	10 mA/g-first cycle; 30 mA/g-20th cycle; 60 mA/g-40th cycle	Sharma et al. (2005)
Molten salt	SnCl$_2$·2H$_2$O and Bat(OH)$_2$ – KOH as alkali flux	Non-uniform aggregates of spherical, rod-shaped with particles size 100–300 nm	1,293-first cycle; 252-50th cycle; 1,077-first cycle; 289-50th cycle	0.005–1; 0.005–3	60 mA/g	Nithyadharseni et al. (2016)
Molten salt	SnCl$_2$·2H$_2$O and BaCl$_2$ – KOH as alkali flux	Aggregated oval-shaped particles size <100 nm in layer-like form	846-first cycle; 183-50th cycle; 914-first cycle; 255-50th cycle	0.005–1; 0.005–3	60 mA/g	Nithyadharseni et al. (2016)
Hydrothermal	Na$_2$SnO$_3$ and Ba(NO$_3$)$_2$	Columnar shape 20–30 μm diameter, 50–100 μm length	~870-first cycle; ~150-50th cycle; ~75-100th cycle	0.005–2.5	60 mA/g	Cha and Kim (2020)
Hydrothermal	Na$_2$SnO$_3$·3H$_2$O and BaCl$_2$·2H$_2$O	Hollow rods 200 nm diameter, 1.5–2 μm length	~700-first cycle; ~125-50th cycle; ~100-100th cycle	0.005–2.5	60 mA/g	Cha and Kim (2020)
Purchased commercially and used	–	Spherical shape 2–3 μm diameter	~675-first cycle; ~125-50th cycle; ~90-100th cycle	0.005–2.5	60 mA/g	Cha and Kim (2020)

BaCO$_3$, barium carbonate; Ba(OH)$_2$, barium hydroxide; KOH, potassium hydroxide; BaCl$_2$, barium chloride.

100 mA/g, a capacity of ~920 mAh/g was obtained at the 10th cycle (Cao et al., 2014). Once again, for enhanced LIB performance, nanocomposites of $CoSnO_3$, carbon nanotubes, and graphene gave better results than nano-$CoSnO_3$ (Zhao et al., 2017). Meanwhile, $CoSnO_3$ nanoboxes (~230 nm particle size) surrounded by RGO pieces also showed high cycling and rate performance (Huang et al., 2017). Stable cycling for 400 cycles has been achieved by coating $CoSnO_3$ nanoboxes with nanocarbon (Wang et al., 2013).

It should be clarified here that the Sn component of $CoSnO_3$, $CaSnO_3$, $SrSnO_3$, and $BaSnO_3$ plays an active role in the electrochemical process, while Co, Ca, Sr, and Ba are inactive and serve as safeguards in alleviating volume expansion (Sharma et al., 2005). Sharma et al. (2005) investigated the electrochemical performance among sol-gel-derived $CaSnO_3$, $SrSnO_3$, and $BaSnO_3$ at different current rates (10 and 60 mA/g). At both rates, the capacity values decreased in the order of $CaSnO_3 > SrSnO_3 > BaSnO_3$. The authors attributed the superior performance of calcium stannate to the lighter weight and smaller radius of calcium (Ca) than the other two. According to the literature (Sharma et al., 2002, 2005; Zhao et al., 2010; Li et al., 2011; Hu et al., 2012), the electrochemical reactions that occur during discharge are as follows:

- $ASnO_3 + 4Li^+ + 4e^- \rightarrow AO + 2Li_2O + Sn$ (A = Co, Ca, Sr, Ba)
- $Sn + xLi^+ + xe^- \leftrightarrow Li_xSn$ (0 < x < 4.4)

The first reaction, which is irreversible, indicates the crystal structure's demolition and results in the creation of lithium oxide (Li_2O) and tin metal implanted in AO. The second process is reversible, showing the formation of a lithium-tin alloy.

7.3.6 Cadmium Metastannate ($CdSnO_3$)

Unlike amorphous $CoSnO_3$, $CdSnO_3$ is crystalline, with a hexagonal unit cell (lattice constants $a = 5.43$ Å, $c = 14.95$ Å) (Sharma et al., 2009). In contrast to the above-mentioned stannates, both tin and cadmium of $CdSnO_3$ are electrochemically active. This indicates that a lithium-cadmium (Li_3Cd) alloy is also formed, in addition to the LiSn alloy, involving the participation of 3 mol of Li^+ ions and resulting in a higher capacity (Sharma et al., 2009; Wang et al., 2014a). It has been reported that $CdSnO_3$ delivered the first and 40th capacities at 1,330 mAh/g and ~475 mAh/g under a current density of 60 mA/g in the 0.005–1.0 V potential range (Sharma et al., 2009). Under a constant current rate of 70 mA/g in the 0.005–1.0 V voltage window, $CdSnO_3$ nanoparticles exhibited specific capacities of 1,546 and 515 mAh/g at the first and 40th cycles, respectively (Wang et al., 2014a). Its capacity was 506 mAh/g when the current density was 150 mA/g (Wang et al., 2014a). The authors believe that the hydrothermal-derived $CdSnO_3$ structure was porous, enabling it to accommodate volume changes while retaining structural stability upon cycling, thus inhibiting capacity degradation (Wang et al., 2014a). It is noticeable that in the past five years, there have not been many works done on $CdSnO_3$, $SrSnO_3$, $CoSnO_3$, and $BaSnO_3$ as anodes in LIB.

7.3.7   ZINC METASTANNATE (ZNSNO$_3$)

Crystalline ZnSnO$_3$, forming nanocubes and nanosheets, was examined for its electrochemical performance (Chen et al., 2012). ZnSnO$_3$ sheets displayed less capacity fading than cubic ZnSnO$_3$, even though both had almost similar first specific capacity. In contrast, the performance of amorphous ZnSnO$_3$, having hollow and solid microcubes, was compared by Xie et al. (2014). The former exhibited higher initial charge and discharge capacities, as well as a reversible capacity over the 50th cycle at a current density of 100 mA/g. Researchers also coated carbon onto the hollow ZnSnO$_3$ microcubes, resulting in larger capacity values over 50 continuous cycles (Xie et al., 2014). The carbon-coated hollow ZnSnO$_3$ also showed superior rate capability compared to the other two. ZnSnO$_3$ nanofibers, with and without carbon, were successfully prepared via the electrospinning technique (Wei et al., 2021). ZnSnO$_3$-carbon nanofibers were obtained by calcination at 873 K for 4 hours in an argon gas condition, whereas nanofibrous pure ZnSnO$_3$ was sintered at 723 K for one day in an air atmosphere. The former displayed a much higher capacity of around 586 mAh/g after 100 cycles than the latter at a current density of 100 mA/g (Wei et al., 2021). Very fine nanoparticles of ZnSnO$_3$ were inserted into RGO sheets and provided a steady capacity over 100 cycles at a current density of 100 mA/g (Ma et al., 2018). Recently, it has been reported that a 3D framework comprised of nanofibers of ZnSnO$_3$ and carbon recorded a good capacity of 695 mAh/g after 200 cycles at a constant current of 100 mA/g (Wang et al., 2023). Another 3D structure of ZnSnO$_3$ and TiO$_2$ nanocomposites documented the first capacity of 1,590 mAh/g, and the capacity was maintained at 780 mAh/g after 200 cycles at a current density of 100 mA/g (Qin et al., 2015). An anode made from ZnSnO$_3$ only performed below par compared to the ZnSnO$_3$-TiO$_2$ nanomaterial (Qin et al., 2015).

7.3.8   ZINC STANNATE (ZN$_2$SNO$_4$)

There are several reports on crystalline Zn$_2$SnO$_4$ in the literature. In this chapter, we have selected a few for discussion. Similar to CdSnO$_3$, zinc in Zn$_2$SnO$_4$ and ZnSnO$_3$, along with tin, are active elements that produce respective alloys with lithium. For solid-state-derived Zn$_2$SnO$_4$, the irreversible and reversible capacities were recorded at 1,150 and 550 mAh/g in the voltage window 0.02–1.5 V under a steady current of 0.12 mA (Belliard et al., 2000). Another solid-state-prepared Zn$_2$SnO$_4$ achieved a first specific capacity of 2,054 mAh/g and ~690 mAh/g capacity upon 50 cycles (current density: 50 mA/g, potential range: 0.05–3 V) (Hou et al., 2010). On the other hand, hydrothermally procured Zn$_2$SnO$_4$ exhibited capacities of 1,384, 988, and 580 mAh/g at the first, second, and 50th cycles, respectively, under a 100 mA/g constant rate in the 0.05–3.0 V voltage (Rong et al., 2006). The shape of Zn$_2$SnO$_4$ was cubic. Another cube-shaped Zn$_2$SnO$_4$ demonstrated a first cycle capacity of ca. 1,700 mAh/g and a 50th cycle capacity of ~366 mAh/g at 300 mA/g current density in the 0.01–2.0 V voltage range (Wang et al., 2014b). However, in the same paper, flower-shaped Zn$_2$SnO$_4$ gave better performance (capacities of 1,750 and 501 mAh/g at the first and 50th cycles, respectively). To improve electrochemical performance, the same authors surrounded Zn$_2$SnO$_4$ nanoflowers with a graphene matrix that could produce a 50th cycle capacity of 540 mAh/g with an initial capacity of 1,967 mAh/g under the same conditions (Wang et al., 2014c). Another group of researchers prepared composites of Zn_2SnO_4 with carbon, displaying ~564 mAh/g capacity at the 40th cycle under a current density of 60 mA/g in the 0.05–3.0 V voltage range, with an initial capacity of ~1,437 mAh/g (Yuan et al., 2010). A lower capacity (~405 mAh/g) was obtained for pure Zn_2SnO_4 after the 40th cycle, even though it initially had a higher capacity of 1,634 mAh/g (Yuan et al., 2010).

In another study, a comparison was made between cubes and hollow boxes of nano-Zn_2SnO_4, finding that the latter exhibited higher rate capability and better capacity retention (Zhao et al., 2013a). In an unrelated report, fast capacity decline was detected in nano-Zn_2SnO_4 in plate form, while nanowires of Zn_2SnO_4 revealed a rather stable capacity during initial cycling. However, the outcome was reversed after cycling for 50 times. Zhu and co-workers (2009) reported that hydrothermal-derived Zn_2SnO_4 displayed a capacity of ~645 mAh/g upon 20 cycles, with initial capacities of ~1,046 and ~1,904 mAh/g during charging and discharging. Song et al. (2013), on the other hand, implanted Zn_2SnO_4 nanocrystals into GNSs to enhance performance. Another approach to improvement is coating Zn_2SnO_4 cubes with PANI, which verified that the capacity increased from 366 for pure Zn_2SnO_4 cubes to 491 mAh/g with PANI at the 50th cycle under a 600 mA/g constant current rate in the 0.01–2.0 V voltage (Zhao et al., 2013a). In addition, a higher initial specific capacity was obtained for that with PANI. The same effect was observed when coating hollow Zn_2SnO_4 boxes with PPy and N-doped carbon, as reported by Wang et al. (2014d) and Zhao et al. (2014).

7.4 OTHER STANNATES

Connor and Irvine (2001) investigated the potential use of magnesium tin oxide (Mg_2SnO_4), cobalt tin oxide (Co_2SnO_4), and manganese tin oxide (Mn_2SnO_4) as anode materials for Li-ion batteries. Initial findings from galvanostatic cycling for 1.5 cycles in the 0.02–1.5 V voltage window revealed that the best reversibility decreased in the order of $Mn_2SnO_4 > Co_2SnO_4 > Mg_2SnO_4$. Mn_2SnO_4 nanoparticles delivered an initial capacity of 1,320 mAh/g in the potential window of 0.01–2.0 V at a current density of 0.067 mA/cm^2 (Lei et al., 2009). Despite the high capacity achieved in the first cycle, Mn_2SnO_4 suffered from significant capacity degradation, producing a capacity of only ~503 mAh/g at the second cycle. The capacity further declined to ~150 mAh/g after the 100th cycle (Lei et al., 2009). Zhu and co-workers (2010) prepared an amorphous tin–manganese oxide film via electrostatic spray deposition, which exhibited a 3D reticular structure. The tin–manganese oxide film exhibited an initial discharge capacity of 1188.3 mAh/g and showed good capacity retention of 656.2 mAh/g up to the 30th cycle. On the other hand, Xiao et al. (2009) prepared SnO_2/Mg_2SnO_4 from hydrothermal-derived $MgSn(OH)_6$ by calcination at 850°C for 5 hours. This SnO_2/Mg_2SnO_4 anode had a high initial capacity of 1,245 mAh/g but poor cyclability, with a reversible capacity of only 350 mAh/g at the end of the 20th cycle. In our opinion, although magnesium and manganese are eco-friendly and low cost, there is not much attention on Mg_2SnO_4 and Mn_2SnO_4 due to their poor performance.

Co_2SnO_4 prepared by the hydrothermal route displayed high capacity and good capacity retention, with an initial capacity of 1533.1 and 555.9 mAh/g at the 50th cycle under a 30 mA/g current density in the 0.005–3.0 V voltage window (Wang et al., 2009b). On the other hand, solid-state-derived Co_2SnO_4 only achieved capacities of 1400.9 mAh/g (first cycle) and 112.8 mAh/g (50th cycle) under the same conditions (Wang et al., 2009b). Unlike its counterpart $CoSnO_3$, it is thought that both cobalt and tin are active electrochemically (Connor and Irvine, 2001; Wang et al., 2009b). To enhance performance, Co_2SnO_4 was coated with carbon (Yuvaraj et al., 2014). The first capacity of the uncoated and carbon-coated Co_2SnO_4 was 1,854 and 1,868 mAh/g, respectively. After the 20th cycle, their capacities declined to 777 mAh/g for the former and 780 mAh/g for the latter at a current density of 40 mA/g in the voltage window of 0–3.0 V. Although not much improvement was observed, the carbon-coated Co_2SnO_4 showed lower charge-transfer resistance and a higher lithium-ion diffusion coefficient based on impedance studies than the uncoated one.

Kishore and co-workers (2004) prepared lithium iron tin oxide, lithium stannoferrite ($LiFeSnO_4$), and lithium indium tin oxide ($LiInSnO_4$) composites using the solid-state reaction method. The starting materials included Li_2CO_3, SnO_2, iron (III) oxide (Fe_2O_3), and indium (III) oxide (In_2O_3). Both compounds are indexed to an orthorhombic unit cell, with lattice parameters for $LiFeSnO_4$ as $a = 3.06$ Å, $b = 5.06$ Å, and $c = 9.85$ Å, and for $LiInSnO_4$ as $a = 3.18$ Å, $b = 5.13$ Å, and $c = 10.17$ Å. An initial capacity of 1,175 mAh/g was achieved by the $Li//LiFeSnO_4$ cell, which gradually decreased to ~105 mAh/g at the 25th cycle in the voltage range from 0.05 to 2.0 V. In contrast, the $Li//LiInSnO_4$ cell delivered a capacity of 1,125 mAh/g in the first cycle, dropping drastically to ~0 mAh/g at the end of the 10th cycle in the voltage window between 0.05 and 2.0 V. Although $LiInSnO_4$ exhibited poor electrochemical performance, it is noteworthy that it was synthesized for the very first time, contributing to the discovery of novel materials in the fields of materials science and crystallography. The authors also attempted to synthesize $LiMSnO_4$, where M = Mn, Al, Ga, Cr, and Ni, but were unsuccessful, as their X-ray patterns exhibited main peaks of SnO_2 instead.

Lithium manganese tin oxide (Li_2MnSnO_4) was synthesized via an oxalyl dihydrazide-assisted combustion method by Mani et al. (2014). Li_2MnSnO_4 was also coated with carbon. Interestingly, the X-ray diffractogram of the carbon-coated Li_2MnSnO_4 calcined at 700°C under an argon atmosphere showed peaks ascribed to hybrid $MnSnO_4$ and Li_2SnO_3 phases instead of Li_2MnSnO_4. A Li-ion half-cell with the uncoated and carbon-coated Li_2MnSnO_4 delivered an initial capacity of ~1,300 mAh/g and ~800 mAh/g, respectively, at a 100 mA/g current density in the 0.05–3 V voltage range. After the 30th cycle, a capacity of ~150 mAh/g was obtained by the Li_2MnSnO_4 anode, while carbon-coated Li_2MnSnO_4 exhibited a capacity of ~600 mAh/g (Mani et al., 2014). As far as we observe, no other literature on Li_2MnSnO_4 compound can be found, similar to $LiMnSnO_4$ and $LiFeSnO_4$. Another anode material tested is tin antimony oxide, $Sn_{0.918}Sb_{0.109}O_2$ (Xiao et al., 2022), prepared via the hydrothermal route, with graphene attached before employing it in LIB. At a 0.1 mA/g current density, the capacities were ~1,176 and 648 mAh/g for the first and 365th cycles, respectively (Xiao et al., 2022). These initial reports indicate that researchers are constantly exploring alternative electrode materials in LIB.

7.5 SUMMARY

In a summary, tin-containing compounds have demonstrated encouraging results and show potential as anodes in LIB. Despite facing a major obstacle—namely, suffering from significant volume expansion—we remain optimistic that this hurdle can be overcome, ultimately improving their performance. We also believe that the potential of tin-based materials is yet to be fully realized to its maximum extent. There is plenty of room for improvement, and much work can still be done to achieve this. Hence, more developments on these materials are foreseen, and we believe they have a bright future in being employed in the next generation of LIBs.

ACKNOWLEDGMENTS

This work was supported by the Malaysian Ministry of Science, Technology and Innovation (MOSTI) under the Nanofund Grant (No. 53-02-03-1089).

REFERENCES

Belliard, F., Connor, P.A., Irvine, J.T.S. (2000). Novel tin oxide-based anodes for Li-ion batteries. *Solid State Ionics* 135, 163–167

Belliard, F., Irvine, J.T.S. (2001). Electrochemical comparison between SnO_2 and Li_2SnO_3 synthesized at high and low temperatures. *Ionics* 7, 16–21

Cao, Y., Zhang, L., Tao, D., Huo, D., Su, K. (2014) Facile synthesis of $CoSnO_3$/graphene nanohybrid with superior lithium storage capability. *Electrochimica Acta* 132, 483-489

Cha, Y.L., Kim, S.H. (2020). Morphological effect on electrochemical properties of $BaSnO_3$ as an anode material for lithium ion batteries. *Journal of Nanoscience and Nanotechnology* 20, 5498–5501

Chandra Bose, A., Kalpana, D., Thangadurai, P., Ramasamy, S. (2002). Synthesis and characterization of nanocrystalline SnO_2 and fabrication of lithium cell using nano-SnO_2. *Journal of Power Sources* 107, 138–141

Chen, J. (2013). Recent progress in advanced materials for lithium ion batteries. *Materials* 6, 156–183

Chen, J., Yano, K. (2013). Highly monodispersed tin oxide/mesoporous starbust carbon composite as high-performance li-ion battery anode. *ACS Applied Materials and Interfaces* 5, 7682–7687

Chen, J.S., Lou, X.W. (2012). SnO_2 and TiO_2 nanosheets for lithium-ion batteries. *Materials Today* 15, 246–254

Chen, J.S., Lou, X.W. (2013). SnO_2-based nanomaterials: Synthesis and application in lithium-ion batteries. *Small* 9, 1877–1893

Chen, S.S., Qin, X. (2014). Tin oxide-titanium oxide/graphene composited as anode materials for lithium-ion batteries. *Journal of Solid State Electrochemistry* 18, 2893–2902

Chen, W., Maloney, S., Wang, W. (2016). Three-dimensional SnO_2/carbon on Cu foam for high-performance lithium ion battery anodes. *Nanotechnology* 27, Article 415401

Chen, Y., Qu, B., Mei, L., Lei, D., Chen, L., Li, Q., Wang, T. (2012). Synthesis of $ZnSnO_3$ mesocrystals from regular cube-like to sheet-like structures and their comparative electrochemical properties in Li-ion batteries. *Journal of Materials Chemistry* 22, 25373–25379

Cherian, C.T., Zheng, M., Reddy, M.V., Chowdari, B.V.R., Sow, C.H. (2013). Zn_2SnO_4 nanowires versus nanoplates: Electrochemical performance and morphological evolution during Li-cycling. *ACS Applied Materials and Interfaces* 5, 6054–6060

Connor, P.A. and Irvine, J.T.S. (2001) Novel tin oxide spinel-based anodes for Li-ion batteries. *Journal of Power Sources* 97-98, 223-225

Courtney, I.A., Dahn, J.R. (1997). Electrochemical and in situ X-ray diffraction studies of the reaction of lithium with tin oxide composites. *Journal of The Electrochemical Society* 144, 2045–2052

Cui, D., Zheng, Z., Peng, X., Li, T., Sun, T., Yuan, L. (2017). Fluorine-doped SnO_2 nanoparticles anchored on reduced graphene oxide as a high-performance lithium ion battery anode. *Journal of Power Sources* 362, 20–26

Cui, L., Shen, J., Cheng, F., Tao, Z., Chen, J. (2011). SnO_2 nanoparticles@polypyrrole nanowires composite as anode materials for rechargeable lithium-ion batteries. *Journal of Power Sources* 196, 2195–2201

Dai, L., Zhong, X., Zou, J., Fu, B., Su, Y., Ren, C., Wang, J., Zhong, G. (2021) Highly ordered SnO_2 nanopillar array as binder-free anodes for long-life and high-rate Li-ion batteries. *Nanomaterials*, 11, 1307

Dang, H.X., Klavetter, K.C., Meyerson, M.L., Heller, A., Mullins, C.B. (2015). Tin microparticles for a lithium ion battery anode with enhanced cycling stability and efficiency derived from Se-doping. *Journal of Materials Chemistry A* 3, 13500–13506

Duan, Y.-K.; Li, Z.-W.; Zhang, S.-C.; Su, T.; Zhang, Z.-H.; Jiao, A.-J.; Fu, Z.-H. (2023) Stannate-based materials as anodes in lithium-ion and sodium-ion batteries: A review. *Molecules*, 28, 5037.

Fan, Q., Chupas, P.J., Whittingham, M.S. (2007) Characterization of amorphous and crystalline tin-cobalt anodes. *Electrochemistry Solid-State Letters*, 10, A274–A278

Faramarzi, M.S., Abnavi, A., Ghasemi, S., Sanaee, Z. (2018) Nanoribbons of SnO_2 as high performance Li-ion anode material. *Materials Research Express*, 5, 065040

Feng, Y., Wu, K., Sun, Y., Guo, Z., Ke, J., Huang, X., Bai, C., Dong, H., Xiong, D., He, M. (2020) Mo-doped SnO_2 nanoparticles embedded in ultrathin graphite nanosheets as a high-reversible capacity, superior-rate and long-cycle-life anode material for lithium-ion batteries. *Langmuir*, 36, 9276–9283

Fu, Z., Liu, P., Ma, J., Guo, B., Chen, X., Zhang, H. (2016) Microwave dielectric properties of low-fired Li_2SnO_3 ceramics co-doped with MgO-LiF. *Materials Research Bulletin*, 77, 78–83

García-Tecedor, M., Bartolomé, J., Maestre, D., Trampert, A., Cremades, A. (2019). Li_2SnO_3 branched nano- and microstructures with intense and broadband white-light emission. *Nano Research* 12, 441–448

Goriparti, S., Miele, E., Angelis, F.D., Fabrizio, E.D., Zaccaria, R.P., Capiglia, C. (2014) Review on recent progress of nanostructured anode materials for Li-ion batteries. *Journal of Power Sources* 257, 421–443

Guan, C., Wang, X., Zhang, Q., Fan, Z., Zhang, H., Fan, H.J. (2014). Highly stable and reversible lithium storage in SnO_2 nanowires surface coated with a uniform hollow shell by atomic layer deposition. *Nano Letters* 14, 4852–4858

Guo, X.W., Fang, X.P., Sun, Y., Shen, L.Y., Wang, Z.X., Chen, L.Q. (2013) Lithium storage in carbon-coated SnO_2 by conversion reaction. *Journal of Power Sources*, 226, 75–81

Hameed, M.U., Dar, S.U., Ali, S., Liu, S., Akram, R., Wu, Z., Butler, I.S. (2017) Facile synthesis of low-dimensional SnO_2 nanostructures: An investigation of their performance and mechanism as anode materials for lithium-ion batteries. *Physica E*, 91, 119–127

Hodeau, J.L., Marezio, M., Santoro, A., Roth, R.S. (1982) Neutron profile refinement of the structures of Li_2SnO_3 and Li_2ZrO_3. *Journal of Solid State Chemistry* 45, 170–179

Hong, Y.J. and Kang, Y.C. (2015) One-pot synthesis of core-shell-structured tin oxide-carbon composite powders by spray pyrolysis for use as anode materials in Li-ion batteries. *Carbon*, 88, 262–269

Hou, X., Cheng, Q., Bai, Y., Zhang, W.F. (2010) Preparation and electrochemical characterization of Zn_2SnO_4 as anode materials for lithium ion batteries. *Solid State Ionics*, 181, 631–634

Hu, G., Zhang, M., Wu, L., Peng, Z., Du, K., Cao, Y. (2016) Enhanced electrochemical performance of $LiNi_{0.5}Co_{0.2}Mn_{0.3}O_2$ cathodes produced via nanoscale coating of Li+-conductive Li_2SnO_3. *Electrochimica Acta*, 213, 547–556

Hu, R.Z., Liu, H., Zeng, M.Q., Liu, J.W., Zhu, M. (2012) Progress on Sn-based thin-film anode materials for lithium-ion batteries. *Chinese Science Bulletin* 57, 4119–4130

Hu, X., Tang, Y., Xiao, T., Jiang, J., Jia, Z., Li, D., Li, B., Luo, L. (2010). Rapid synthesis of single-crystalline SrSn(OH)6 nanowires and the performance of $SrSnO_3$ nanorods used as anode materials for li-ion battery. *Journal of Physical Chemistry C* 114, 947–952

Hu, Y., Xu, K., Kong, L., Jiang, H., Zhang, L., Li, C. (2014) Flame synthesis of single crystalline SnO nanoplatelets for lithium-ion batteries. *Chemical Engineering Journal*, 242, 220–225

Huang, B., Yang, J., Zhou, X. (2014). Hierarchical SnO_2 with double carbon coating composites as anode materials for lithium ion batteries. *Journal of Solid State Electrochemistry* 18, 2443–2449

Huang, F., Yuan, Z., Zhan, H., Zhou, Y., Sun, J. (2003) A novel tin-based nocomposite oxide as negative-electrode materials for Li-ion batteries. *Materials Letters* 57, 3341-3345

Huang, J., Ma, Y., Xie, Q., Zheng, H., Yang, J., Wang, L., & Peng, D.-L. (2017). 3D graphene encapsulated hollow $CoSnO_3$ nanoboxes as a high initial coulombic efficiency and lithium storage capacity anode. *Small*, 14(10), 1703513.

Huang, Y., Wang, Q., Wang, Y. (2012) Preparation and electrochemical characterisation of polypyrrole-Li_2SnO_3 anode materials for lithium-ion batteries. *Micro & Nano Letters* 7, 1278–1281

Idota, Y., Kubota, T., Matsufuji, A., Maekawa, Y., Miyasaka, T. (1997) Tin-based amorphous oxide: A high-capacity lithium-ion-storage material. *Science* 276, 1395–1419

Iqbal, M.Z., Wang, F., Zhao, H., Rafique, M.Y., Wang, J., Li, Q. (2012) Structural and electrochemical properties of SnO nanoflowers as an anode material for lithium ion batteries. *Scripta Materialia*, 67, 665–668

Jaśkaniec, S., Kavanagh, S.R., Coelho, J., Ryan, S., Hobbs, C., Walsh, A., Scanlon, D.O., Nicolosi, V. (2021). Solvent engineered synthesis of layered SnO for high-performance anodes. *Npj 2D Materials and Applications, 5.* 27 (9 pp)

Kamali, A.R., Fray, D.J. (2011). Tin-based materials as advanced anode materials for lithium ion batteries: A review. *Reviews on Advanced Materials Science* 27, 14–24

Kepler, K.D., Vaughey, J.T., Thackeray, M.M. (1999) Copper-tin anodes for rechargeable lithium batteries: an example of the matrix effect in an intermetallic system. *Journal of Power Sources* 81–82, 383–387

Kim, H.S., Park, S.-S., Kang, S.H. Kang, Sung, Y.-E. (2014) Effect of particle shape on the electrochemical properties of $CaSnO_3$ as an anode material for lithium-ion batteries. *Journal of Applied Electrochemistry* 44, 789–796

Kim, I.-S., Blomgren, G.E., Kumta, P.N. (2004) Sn/C composite anodes for Li-ion batteries. *Electrochemistry Solid-State Letters*, 7, A44–A48.

Kishore, M. V. V. M.s., Varadaraju, U. V., Raveau, B. (2004) Electrochemical performance of LiMSnO4 (M = Fe, In) phases with ramsdellite structure as anodes for lithium batteries. *Journal of Solid State Chemistry*, 177(11), 3981–3986

Lang, G. (1966) Strukturvergleiche an ternären und quarternären Oxiden. *Zeitschrift für anorganische und allgemeine Chemie* 348, 246–256

Lei, S., Tang, K., Chen, C., Jin, Y., Zhou, L. (2009) Preparation of Mn_2SnO_4 nanoparticles as the anode material for lithium secondary battery. *Materials Research Bulletin* 44, 393–397

Li, C., Zhu, Y., Fang, S., Wang, H., Gui, Y., Bi, L., Chen, R. (2011) Preparation and characterization of $SrSnO_3$ nanorods. *Journal of Physics and Chemistry of Solids* 72, 869-874

Li, L., Peng, S., Cheah, Y.L., Wang, J., Teh, P., Ko, Y., Wong C., Srinivasan, M. (2013) Electrospun eggroll-like $CaSnO_3$ nanotubes with high lithium storage performance. *Nanoscale* 5, 134–138

Lin, Y., Wu, J., Chen, W. (2013) Enhanced electrochemical performance of LiFePO$_4$/C prepared by sol-gel synthesis with dry ball-milling. *Ionics* 19, 227–234

Liu, H., Huang, J., Li, X., Liu, J., Zhang, Y. (2012) SnO$_2$ nanorods grown on graphite as high-capacity anode material for lithium ion batteries. *Ceramics International*, 38, 5145–5149

Liu, H., Huang, J., Li, X., Liu, J., Zhang, Y. (2013) Green synthesis of SnO$_2$ nanosheets and their electrochemical properties. *Ceramics International* 39, 3413–3415

Liu, L., An, M., Yang, P., Zhang, J. (2015). Superior cycle performance and high reversible capacity of SnO$_2$/graphene composite as an anode material for lithium-ion batteries. *Scientific Reports* 5

Liu, X.-Y., Han, Y.-L., Li, Q., Pan, D. (2017) SnO$_2$ nanoparticles for lithium-ion batteries prepared by sol-gel method. *Key Engineering Materials*, 727, 719–725

Ma, Y., Jiang, R., Li, D., Dong, Y., Liu, Y., Zhang, J. (2018) Embedding ultrafine ZnSnO$_3$ nanoparticles into reduced graphene oxide composites as high-performance electrodes for lithium ion batteries. *Nanotechnology* 29, 195401

Macías, M.A., Martínez, J.A.H., Gauthier, G.H., Rodriguez, J.E., Avila, H., Pinto, J., Pinilla, J. (2011) Structural and microstructural characterization of tin(II) oxide useful as anode material in lithium rechargeable batteries obtained from a different synthesis route at room temperature. *Materials Research*, 14, 172–177

Mani, V., Babu, G., Kalaiselvi, N. (2014) Li2MnSnO$_4$/C anode for high capacity and high rate lithium battery applications. *Electrochimica Acta* 133, 347–353

Manthiram. A. (2011) Materials challenges and opportunities of lithium ion batteries. *Journal of Physical Chemistry Letters* 2, 176–184

Meduri, P., Pendyala, C., Kumar, V., Sumanasekera, G.U., Sunkara, M.K. (2009) Hybrid tin oxide nanowires as stable and high capacity anodes for Li-ion batteries. *Nano Letters* 9, 612–616

Mei, L., Li, C., Qu, B., Zhang, M., Xu, C., Lei, D., Chen, Y., Xu, Z., Chen, L., Li, Q., Wang, T. (2012) Small quantities of cobalt deposited on tin oxide as anode material to improve performance of lithium-ion batteries. *Nanoscale*, 4, 5731–5737

Morimoto, H. and Tobishima, S.I. (2005) Anode behavior of electroplated rough surface Sn thin films for lithium-ion batteries. *Journal of Power Sources*, 146, 469–472

Mou, H., Xiao, W., Miao, C., Li, R., Yu, L. (2020). Tin and tin compound materials as anodes in lithium-ion and sodium-ion batteries: A review. *Frontiers in Chemistry*, 8, Article 141

Mousavi, M.R., Abolhassani, R., Hosseini, M., Akbarnejad, E., Mojallal, M.H., Ghasemi, S., Mohajerzadeh, S., Sanaee, Z. (2021). Antimony doped SnO$_2$ nanowire@C core-shell structure as a high-performance anode material for lithium-ion battery. *Nanotechnology* 32

Nithyadharseni, P., Reddy, M.V., Ozoemena, K.I., Ezema, F.I., Balakrishna, R.G., R. Chowdari, B.V. (2016) Electrochemical performance of BaSnO$_3$ anode material for lithium-ion battery prepared by molten salt method. *Journal of The Electrochemical Society*, 163 (3) A540–A545

Palaniyandy, N., Kebede, M.A., Ozoemena, K.I., Mathe, M.K. (2019) Rapidly microwave-synthesized SnO2 nanorods anchored on onion-like carbons (OLCs) as anode material for lithium-ion batteries. *Electrocatalysis* 10, 314–322

Park, M.-S., Kang, Y.-M., Wang, G.-X., Dou, S.-X., Liu, H.-K. (2008) The effect of morphological modification on the electrochemical properties of SnO2 nanomaterials. *Advanced Functional Materials* 18, 455–461

Park, M.S., Wang, G.X., Kang, Y.M., Wexler, D., Dou, S.X., Liu, H.K. (2007). Preparation and electrochemical properties of SnO$_2$ nanowires for application in lithium-ion batteries. *Angewandte Chemie-International Edition* 46, 750–753

Qin, Y.L., Zhang, F.F., Du, X.C., Huang, G., Liu, Y.C., Wang, L.M. (2015). Controllable synthesis of cube-like ZnSnO$_3$@TiO$_2$ nanostructures as lithium ion battery anodes. *Journal of Materials Chemistry A* 3, 2985–2990

Rong, A., Gao, X.P., Li, G. R., Yan, T. Y., Zhu, H. Y., Qu, J. Q., Song, D. Y. (2006) Hydrothermal synthesis of Zn_2SnO_4 as anode materials for Li-ion battery. *Journal of Physical Chemistry B* 110, 14754–14760

Saddique, J., Shen, H., Ge, J., Huo, X., Rahman, N., Mushtaq, M., Althubeiti, K., Al-Shehri, H. (2022) Synthesis and characterization of $Sn/SnO_2/C$ nano-composite structure: high-performance negative electrode for lithium-ion batteries. *Materials* 15, 2475.

Sharma, N., Shaju, K.M., Subba Rao, G.V., Chowdari, B.V.R. (2002) Sol-gel derived nano-crystalline $CaSnO_3$ as high capacity anode material for Li-ion batteries. *Electrochemistry Communications* 4, 947-952

Sharma, N., Shaju, K.M., Subba Rao, G.V., Chowdari, B.V.R. (2005) Anodic behaviour and X-ray photonelectron spectroscopy of ternary tin oxides. *Journal of Power Sources* 139, 250-260

Sharma, Y., Sharma, N., Subba Rao, G.V., Chowdari, B.V.R. (2009) Lithium-storage and cycleability of nano-$CdSnO_3$ as an anode material for lithium-ion batteries. *Journal of Power Sources 192, 627-635*

Shin, N., Kim, M., Ha, J., Kim, Y.-T., Choi, J. (2022) Flexible anodic SnO_2 nanoporous structures uniformly coated with polyaniline as a binder-free anode for lithium ion batteries. *Journal of Electroanalytical Chemistry* 914, 116296

Sivashanmugam, A., Kumar, T.P., Renganathan, N.G., Gopukumara, S., Wohlfahrt-Mehrens, M., Garche, J. (2005) Electrochemical behavior of Sn/SnO_2 mixtures for use as anode in lithium rechargeable batteries. *Journal of Power Sources* 144, 197–203

Song, W., Xie, J., Hu, W., Liu, S., Cao, G., Zhu, T., Zhao, X. (2013) Facile synthesis of layered Zn_2SnO_4/graphene nanohybrid by a one-pot route and its application as high-performance anode for Li-ion batteries. *Journal of Power Sources* 229, 6–11

Sun, S. and Liang, S. (2017). Morphological zinc stannate: synthesis, fundamental properties and applications. *Journal of Materials Chemistry A* 5(39), 20534–20560

Tao, H., Feng, Z., Liu, H., Kan, X., Chen, P. (2011) Reality and future of rechargeable lithium batteries. *The Open Materials Science Journal* 5 (Suppl 1: M2), 204–214

Tarakina, N.V., Denisova, T.A., Maksimova, L.G., Baklanova, Y.V., Tyutyunnik, A.P., Berger, I.F., Zubkov, V.G., van Tendeloo, G. (2009) Investigation of stacking disorder in Li_2SnO_3. *Z Kristallogr* Suppl. 30, 375–380

Thomas, R. and Rao, G.M. (2014) Phase and dimensionality of tin oxide at graphene nanosheet array and its electrochemical performance as anode for lithium ion battery. *Electrochimica Acta* 125, 380–385

Uchiyama, H., Hosono, E., Honma, I., Zhou, H., Imai, H. (2008) A nanoscale meshed electrode of single-crystalline SnO for lithium-ion rechargeable batteries. *Electrochemistry Communications* 10, 52–55

Vaughey, J.T., Geyer, A.M., Fackler, N., Jonhson, C.S., Edstrom, K., Bryngelsson, H., Benedek, R., Thackeray, M.M. (2007) Studies of layered lithium metal oxide anodes in lithium cells. *Journal of Power Sources* 174, 1052–1056

Veerappana, G., Yoo, S., Zhang, K., Ma, M., Kang, B., Park, J.H. (2016) High-reversible capacity of perovskite $BaSnO_3$/rGO composite for lithium-ion battery anodes. *Electrochimica Acta* 214, 31–37

Xiao, T., Tang, Y., Jia, Z., Feng, S. (2009) Synthesis of SnO_2/Mg_2SnO_4 nanoparticles and their electrochemical performance for use in Li-ion battery electrodes. *Electrochimica Acta* 54, 2396-2401

Wakihara, M. (2001) Recent developments in lithium ion batteries. *Materials Science and Engineering* R33, 109–134

Wan, N., Yu, P., Sun, S., Wu, Q., Li, T., Bai, Y. (2014) Copper and nitrogen co-doped SnO_2 hierarchical microspheres as a novel anode material for lithium ion batteries. *Materials Letters* 133, 168–170

Wang, G., Gao, X.P., Shen, P.W. (2009b) Hydrothermal synthesis of Co_2SnO_4 nanocrystals as anode materials for Li-ion batteries. *Journal of Power Sources* 192, 719–723

Wang, J., Zhao, H., Liu, X., Wang, J., Wang, C. (2011) Electrochemical properties of SnO$_2$/carbon composite materials as anode material for lithium-ion batteries. *Electrochimica Acta* 56, 6441–6447

Wang, Q., Huang, Y., Miao, J., Zhao, Y., Wang, Y. (2012a) Synthesis and properties of carbon-doped Li$_2$SnO$_3$ nanocomposite as cathode materials for lithium-ion batteries. *Materials Letters* 71, 66–69

Wang, Q., Huang, Y., Miao, J., Wang, Y., Zhao, Y. (2012b) Hydrothermal derived Li$_2$SnO$_3$/C composite as negative electrode materials for lithium-ion batteries. *Applied Surface Science* 258, 6923–6929

Wang, Q., Huang, Y., Zhao, Y., Wang, Y. (2012c) Synthesis and properties of Li$_2$SnO$_3$/polyaniline nanocomposites as negative electrode material for lithium-ion batteries. *Applied Surface Science* 258, 9896–9901

Wang, Q., Huang, Y., Zhao, Y., Zhang, W., Wang, Y. (2013) Preparation of Li$_2$SnO$_3$ and its application in lithium-ion batteries. *Surface and Interface Analysis* 45, 1297–1303

Wang, L., Zhang, W., Wang, C., Wang, D., Liu, Z., Hao, Q., Wang, Y., Tang, K., Qian, Y. (2014a) A facile synthesis of highly porous CdSnO$_3$ nanoparticles and their enhanced performance in lithium-ion batteries. *Journal of Materials Chemistry A* 2, 4970–4974

Wang, K., Huang, Y., Huang, H., Y. Zhao, Qin, X., Sun, X., Wang, Y. (2014b) Hydrothermal synthesis of flower-like Zn$_2$SnO$_4$ composites and their performance as anode materials for lithium-ion batteries. *Ceramics International* 40, 8021–8025

Wang, K., Huang, Y., Shen, Y., Xue, L., Huang, H., Wu, H., Wang, Y. (2014c) Graphene supported Zn$_2$SnO$_4$ nanoflowers with superior electrochemical performance as lithium-ion battery anode. *Ceramics International* 40, 15183–15190

Wang, K., Huang, Y., Han, T., Zhao, Y., Huang, H., Xue, L. (2014d) Facile synthesis and performance of polypyrrole-coated hollow Zn$_2$SnO$_4$ boxes as anode materials for lithium-ion batteries. *Ceramics International* 40, 2359–2364

Wang, Q., Yang, S., Miao, J., Zhang, Y., Zhang, D., Chen, Y., Li, Z. (2019) Synthesis of graphene supported Li2SiO3/Li$_2$SnO$_3$ anode material for rechargeable lithium ion batteries. *Applied Surface Science* 469, 253–261

Wang, X., Zhao, X., Wang, Y. (2021) A Nanosheet-assembled SnO2-integrated anode. *Molecules* 26, 6108

Wang, X., Zheng, T., Cheng, Y.-J., Yin, S., Xia, Y., Ji, Q., Xu, Z, Liang, S., Ma, L., Zuo, X., Meng, J.-Q., Zhu, J., Müller-Buschbaum, P. (2021) SnO$_2$/Sn/Carbon nanohybrid lithium-ion battery anode with high reversible capacity and excellent cyclic stability. *Nano Select* 2, 642–653

Wang, X.L., Han, W.Q., Chen, J., Graetz, J. (2010). Single-crystal intermetallic M-Sn (M = Fe, Cu, Co, Ni) nanospheres as negative electrodes for lithium-ion batteries. *ACS Applied Materials and Interfaces* 2, 1548–1551

Wang, Y., Djerdj, I., Smarsly, B., Antonietti, M. (2009a) Antimony-doped SnO$_2$ nanopowders with high crystallinity for lithium-ion battery electrode. *Chemistry of Materials* 21, 3202–3209

Wang, Z., Chen, L., Feng, J., Liu, S., Wang, Y., Fan, Q., Zhao, Y. (2019) In-situ grown SnO$_2$ nanospheres on reduced GO nanosheets as advanced anodes for lithium-ion batteries. *ChemistryOpen* 8, 712–718

Wang, Z., Wang, Z., Liu, W., Xiao, W., Lou, X.W. (2013). Amorphous CoSnO$_3$@C nanoboxes with superior lithium storage capability. *Energy and Environmental Science* 6, 87–91

Wang, K., Zhang, S., Hou, Z., Wang, L., An, P., Jia, J., Li, Y., Zhang, P. (2023) Nanofibrous ZnSnO$_3$/C composite derived from natural cellulose substance as an enhanced lithium-ion battery anode. *Materials Letters* 331, 133435

Wei, J.-L., Jin, X.-Y., Yu, M.-C., Wang, L., Guo, Y.-H., Dong, S.-T., Zhang, Y.-M. (2021). Electrospun ZnSnO$_3$/C nanofibers as an anode material for lithium-ion batteries. *Journal of Electronic Materials* 50(8), 4945–4953

Wen, Z., Zheng, F., Yu, H., Jiang, Z., Liu, K. (2013) Hyrothermal synthesis of flower-like SnO_2 nanorods bundles and their application for lithium ion battery. *Materials Characterization* 76, 1–5

Whittingham, M.S. (2008) Materials challenges facing electrical energy storage. *MRS Bulletin* 33, 411–419

Winter, M. and Besenhard, J.O. (1999) Electrochemical lithiation of tin and tin-based interme-tallics and composites. *Electrochimica Acta* 45, 31–50

Wu, H. and Cui, Y. (2012) Designing nanostructured Si anodes for high energy lithium ion batteries. *Nano Today* 7(5), 414–429

Wu, H.B., Chen, J.S., Lou, X.W., Hng, H.H. (2011). Synthesis of SnO_2 hierarchical structures assembled from nanosheets and their lithium storage properties. *Journal of Physical Chemistry C* 115, 24605–24610

Xiao, B., Wu, G., Wang, T., Wei, Z., Sui, Y., Shen, B., Qi, J., Wei, F., Meng, Q., Ren. Y., Xue, X., Zheng, J., Mao, J., Dai, K., Yan, Q. (2022) Tin antimony oxide @graphene as a novel anode material for lithium ion batteries. *Ceramics International* 48(2), 2118–2123

Xie, Q., Ma, Y., Zhang, X., Guo, H., Lu, A., Wang, L., G. Yue, Peng, D.-L. (2014) Synthesis of amorphous $ZnSnO_3$-C hollow microcubes as advanced anode materials for lithium ion batteries. *Electrochimica Acta* 141, 374–383

Yang, S., Zavalij, P.Y., Whittingham, M.S. (2003). Anodes for lithium batteries: Tin revisited. *Electrochemistry Communications* 5, 587–590

Yang, S., Zhang, J., Wang, Q., Miao, J., Zhang, C., Zhao, L., Zhang, Y. (2019) Li_2SiO_3@Li_2SnO_3/SnO_2 as a high performance lithium-ion battery material. *Materials Letters* 234, 375–378

Ye, J., Zhang, H., Yang, R., Li, X., Qi, L. (2010) Morphology-controlled synthesis of SnO_2 nanotubes by using 1D silica mesostructures as sacrificial templates and their applica-tions in lithium-ion batteries. *Small* 6, 296–306

Yin, L., Chai, S., Wang, F., Huang, J., Li, J., Liu, C., Kong, X. (2016) Ultrafine SnO_2 nanoparti-cles as a high performance anode material for lithium ion battery. *Ceramics International* 42, 9433–9437

Yoon, S. and Manthiram, A. (2011) Nanostructured Sn-Ti-C composite anodes for lithium ion batteries. *Electrochimica Acta* 56, 3029–3035

Yu, L., Cai, D., Wang, H., Titirici, M.-M. (2013) Hydrothermal synthesis of SnO_2 and SnO_2@C nanorods and their application as anode materials in lithium-ion batteries. *RSC Advances* 3, 17281–17286

Yuan, J., Chen, C., Hao, Y., Zhang, X., Zou, B., Agrawal, R., Wang, C., Yu, H., Zhu, X., Yu, Y., Xiong, Z., Luo, Y., Li, H., Xie, Y. (2017) SnO_2/polypyrrole hollow spheres with improved cycle stability as lithium-ion battery anodes. *Journal of Alloys and Compounds* 691. 34–39

Yuan, W.S., Tian, Y.W., Liu, G.Q. (2010) Synthesis and electrochemical properties of pure phase Zn2SnO4 and composite Zn_2SnO_4/C. *Journal of Alloys and Compounds* 506, 683–687

Yuvaraj, S., Amaresh, S., Lee, Y. S., Kalai Selvan, R. (2014) Effect of carbon coating on the electrochemical properties of Co_2SnO_4 for negative electrodes in Li-ion batteries. *Royal Society of Chemistry Advances* 4, 6407–6416

Zakuan, N.S., Woo, H.J., Teo, L.P., Kufian, M.Z., Osman, Z. (2019). Li2 SnO_3 anode synthe-sized via simplified hydrothermal route using eco-compatible chemicals for lithium-ion battery. *Journal of The Electrochemical Society* 166, A2341–A2348

Zhang, C.Q., Tu, J.P., Huang, X.H., Yuan, Y.F., Wang, S.F., Mao F. (2008) Preparation and elec-trochemical performances of nanoscale $FeSn_2$ as anode material for lithium ion batteries. *Journal of Alloys and Compounds* 457, 81–85

Zhang, D.W., Zhang, S.Q., Jin, Y., Yi, T.H., Xie, S., Chen, C.H. (2006) Li_2SnO_3 derived sec-ondary Li-Sn alloy electrode for lithium-ion batteries. *Journal of Alloys and Compounds* 415, 229–233

Zhang, H., He, Q., Wei, F., Tan, Y., Jiang, Y., Zheng, G., Ding, G., Jiao, Z. (2014) Ultrathin SnO nanosheets as anode materials for rechargeable lithium-ion batteries. *Materials Letters* 120, 200–203

Zhang, H.-J. Shu, J., Wei, X., Wang, K.-X., Chen, J.-S. (2013) Cerium vanadate nanoparticles as a new anode material for lithium ion batteries. *RSC Advances* 3, 7403–7407

Zhang, R., Whittingham, M.S. (2010) Electrochemical behavior of the amorphous tin-cobalt anode. *Electrochemistry Solid-State Letters* 13, A184–A187

Zhang, W., Feng, L., Chen, H., Zhang, Y. (2019) Hydrothermal synthesis of SnO_2 nanorod as anode materials for lithium-ion battery. *Nano: Brief Reports and Reviews* 14, 1950109 (9 pages)

Zhang, W.-J. (2011) A review of the electrochemical performance of alloy anodes for lithium-ion batteries. *Journal of Power Sources* 196, 13–24

Zhao, S., Bai, Y., Zhang, W.-F. (2010) Electrochemical performance of flowerlike $CaSnO_3$ as high capacity anode material for lithium-ion batteries. *Electrochimica Acta* 55, 3891-3896

Zhao, Q., Ma, L., Zhang, Q., Wang, C., Xu, X. (2015) SnO2-based nanomaterials: synthesis and application in lithium-ion batteries and supercapacitors. *Journal of Nanomaterials* 2015, 850147

Zhao, X., Wang, G., Zhou, Y., Wang, H. (2017) Flexible free-standing ternary $CoSnO_3$/graphene/carbon nanotubes composite papers as anodes for enhanced performance of lithium-ion batteries, *Energy* 118, 172–180

Zhao, Y., Li, X., Yan, B., Li, D., Lawes, S., Sun, X. (2015). Significant impact of 2D graphene nanosheets on large volume change tin-based anodes in lithium-ion batteries: A review. *Journal of Power Sources* 274, 869–884

Zhao, Y., Huang, Y., Wang, Q., Wang, K., Zong, M., Wang, L., Sun, X. (2014) Hollow Zn_2SnO_4 boxes coated with N-doped carbon for advanced lithium-ion batteries, *Ceramics International* 40, 2275–2280

Zhao, Y., Huang, Y., Wang, Q., Wang, K., Zong, M., Wang, L., Zhang, W., Sun, X. (2013a). Preparation of hollow Zn_2SnO_4 boxes for advanced lithium-ion batteries. *RSC Advances* 3, 14480–14485.

Zhao, Y., Huang, Y., Wang, Q., Wang, X., Zong, M., Wu, H., Zhang, W. (2013b) Graphene supported Li_2SnO_3 as anode material for lithium-ion batteries. *Electronic Materials Letters* 9, 683–686

Zhao, Y., Huang, Y., Wang, Q., Zhang, W., Wang, K., Zong, M. (2013c) The study on Li-storage performances of bamboo charcoal (BC) and BC/Li_2SnO_3 composites. *Journal of Applied Electrochemistry* 43, 1243–1248

Zhao, Y., Huang, Y., Xue, L., Sun, X., Wang, Q., Zhang, W., Wang, K., Zong, M. (2013d) Polyaniline (PANI) coated Zn_2SnO_4 cube as anode materials for lithium batteries, *Polymer Testing* 32, 1582–1587

Zhu, X., Zhu, J.,Yao, Y., Zhou, Y., Tang, Y., Wu, P. (2015) Facile template-directed synthesis of carbon-coated SnO_2 nanotubes with enhanced Li-storage capabilities, *Materials Chemistry and Physics* 163, 581–586

Zhu, X.J., Geng, L.M., Zhang, F.Q., Liu, Y.X., Cheng, L.B. (2009) Synthesis and performance of Zn_2SnO_4 as anode materials for lithium ion batteries by hydrothermal method. *Journal of Power Sources* 189, 828–831

Zhu, X.J., Guo, Z.P., Zhang, P., Du, G.D., Poh, C.K., Che, Z.X., Li, S., Liu, H.K. (2010) Three-dimensional reticular tin-manganese oxide composite anode materials for lithium ion batteries. *Electrochimica Acta* 55, 4982–4986

Zoller, F., Böhm, D., Bein, T., Fattakhova-Rohlfing, D. (2019). Tin oxide based nanomaterials and their application as anodes in lithium-ion batteries and beyond. *ChemSusChem* 12, 4140–4159

Zou, Y., Wang, Y. (2013) Microwave solvothermal synthesis of flower-like SnS_2 and SnO_2 nanostructures as high-rate anodes for lithium ion batteries. *Chemical Engineering Journal* 229, 183–189

8 Sustainable Materials for Energy Conversion System

Grishika Arora and Hieng Kiat Jun

A society grows great when old men plant trees whose shade they know they shall never sit in.

– Greek Proverb

8.1 INTRODUCTION

The saying "Sustainable development is development that meets the needs of the present without compromising the ability of future generations to meet their own needs" shows that the idea of sustainable development is not new (from the 1987 publication of the Brundtland Commission Report) (Park, 2018). The so-called "cradle-to-grave" strategy, according to Narayan, mandates that produced goods be conceived and engineered from "conception to reincarnation." It is important to figure out how biomass is used in the context of the entire carbon cycle. The lithosphere (for example, limestone), the biosphere (for example, plants and animals), the hydrosphere (for example, bicarbonate dissolved in the oceans), and the atmosphere (CO_2) are the four main reservoirs of the carbon on the planet known as the carbon cycle. Recent human activities (such as the burning of fossil fuels and extensive deforestation) cause a significant imbalance in the carbon cycle, resulting in a large-scale release of CO_2 into the atmosphere that cannot be entirely offset by photosynthesis or oceanic dissolution. It causes a significant buildup of CO_2 in the atmosphere, which aids in the process of global warming. People are increasingly aware that measures must be taken to lessen CO_2 generation to rebalance the carbon cycle (Song et al., 2009).

Since only a certain number of energy sources (solar, wind, hydro, tidal, and geothermal) are renewable and none of our natural resources are limitless, the implications of sustainable development affect every aspect of human effort. Sustainability, handling resources, sustainable production, technology for renewable and clean energy, and management of water and air are all included in the vast subject of sustainable development. It is the procedure by which scientific analysis ("the triple bottom line") is used to inform strategies for ongoing improvements in the economy, environment, and society (Green et al., 2012).

DOI: 10.1201/9781003314424-8

Sustainable development issues are inherently extremely interconnected as the three parts of the triple bottom line overlap as illustrated in Figure 8.1.

We have found it helpful to use the 2×3 taxonomy of the objectives outlined in the extensive literature that defines or discusses sustainable development in order to clarify the definitional ambiguities connected with it as shown in Table 8.1 (Parris and Kates, 2003; Ness, 2020). There are three main categories listed in the first column under the heading "what is to be sustained": community, health care, and environment. Most of the literature focuses on sustaining life support systems/health care/medical care, in which the ecosystem or nature provides resources for the real-world existence of humanity, such as products and services (Lawton, 1998; Costanza et al., 1997).

However, a substantial portion of literature gives more weight to the beauty and richness of the natural world than to its practical value. Finally, there are assertions that unique and threatened communities should be preserved, including cultural diversity, livelihoods, groups, and locations (Parris and Kates, 2003; Muehlebach, 2001).

Three distinct categories—people, business, and society—should also be created. Most of the early writing was focused on the economy, with its producing sectors providing both the needed labour and the desired consumption and wealth. In this literature, economic growth is presented as offering incentives and methods for investment as well as financial resources for upkeep and restoration of the ecology (Solow, 1993). People are the focus now, with a focus on human development, improved life expectancy, education, equity investment, and prosperity. Furthermore, there are also calls for the development of societies that place a strong emphasis on the stability and well-being of national states, regions, and organizations as well as the social capital of ties within communities (Parris and Kates, 2003; Adeney, 2002).

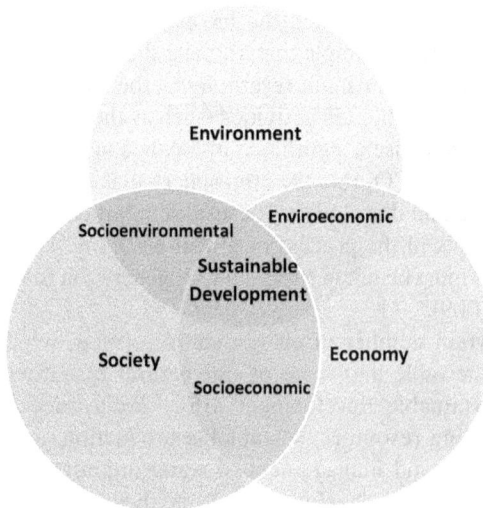

FIGURE 8.1 Sustainable growth focuses on the "triple bottom line" of social, economic, and environmental factors. Sustainable growth happens when all three of them exist side by side. Adapted from Green et al. (2012).

On the other hand, although the connections between sustainable development and materials are clear-cut and fundamental, consumers of cutting-edge technologies frequently fail to notice them. Materials often play a supporting role in technology and are not seen as being crucial by end consumers. For instance, when operating a modern car, the owner appreciates the increased fuel efficiency while frequently being unaware of the material advances in technology (composite and alloys that are lightweight, digital control systems, higher-density rechargeable batteries, etc.) which made gain feasible (Green et al., 2012). Figure 8.2 shows some examples of materials for sustainable development. In order for technologists, policymakers, and business leaders, this chapter intends to emphasize the significant role that materials contribute to environmentally friendly growth and to define and plan the future responsibilities of advanced materials in the sustainable development area. The sustainable development is focused on materials involved in energy conversion systems, especially solar energy devices and energy storage systems.

Perhaps the biggest obstacle to sustainable growth is changing people's behaviour. Regional and global strategies will need to be developed by governing bodies, politicians, and scientists; Apelian covered this in the first article of the bulletin. Technology alone won't be the solution. Additionally, he contrasts the open loop of human civilization with the closed cycle of the ecological system, in which there is no "waste" (LeStar et al., 2012) As the world's population increases, the most crucial issue may not be how many people the planet can sustain, but rather how many people it can support sustainably while maintaining an adequate standard of living (Green et al., 2012). Let's briefly first understand how do some energy conversion and storage systems function.

TABLE 8.1
Sustainable Development Objectives' Taxonomy

What Is to Be Sustained	What Is to Be Developed
Nature	People
Earth	Child survival
Biodiversity	Life expectancy
Ecosystems	Education
	Equity
	Equal opportunity
Life support	Economy
Ecosystem services	Wealth
Resources	Productive sectors
Environment	Consumption
Community	Society
Cultures	Institutions
Groups	States
Places	Regions
	Social capital

Adapted from Parris and Kates (2003) and Ness (2020).

(a)

(b)

FIGURE 8.2 Sustainable development in materials: (a) the apparent solution for all modes of transportation is to make them lighter; (b) "urban mining" includes the recycling of electronic waste, which can contain more important materials than a mine itself. Adapted from Green et al. (2012).

8.2 ENERGY CONVERSION AND STORAGE SYSTEMS

8.2.1 SOLAR ENERGY

Based on the various energy conversion and storage, the integrated energy conversion and storage systems (IECSSs) can be classified into two classes as indicated in Figure 8.3, which are photovoltaic (PV) charging based and photocatalytic charging based. During the photo-charging and photo-discharging operations, the energy conversion and storage sections are typically electrochemical separation. The performance activities of these sections are independent (Figure 8.3a). The PV-charging procedure does not involve any photo-induced redox processes. The solar energy conversion sections of the IECSSs are usually from semiconductor solar cell types, which include but not limited to silicon solar cells, organic solar cells, and perovskite solar cells (PSCs). PV solar cells may fully or partially provide the charging voltage for the energy storage component. In contrast, the method used in a photocatalytic charged arrangement for energy storage (photo-charging) will use photo-induced redox processes as shown in Figure 8.3b. The photoelectrodes hold a significant substance that will be stimulated during the photo-charging procedure to produce the electron hole pairs. The resultant electrons will then move to the energy storage. The electrons originating from the counter electrode will subsequently balance the redox reactions in which both the electrode and the holes participate. The electrolyte that exists between each of these electrodes' cations or anions will serve as a counterweight when the charges from the energy storage electrode return to the counter electrode during the discharge process (Luo and Wang, 2017).

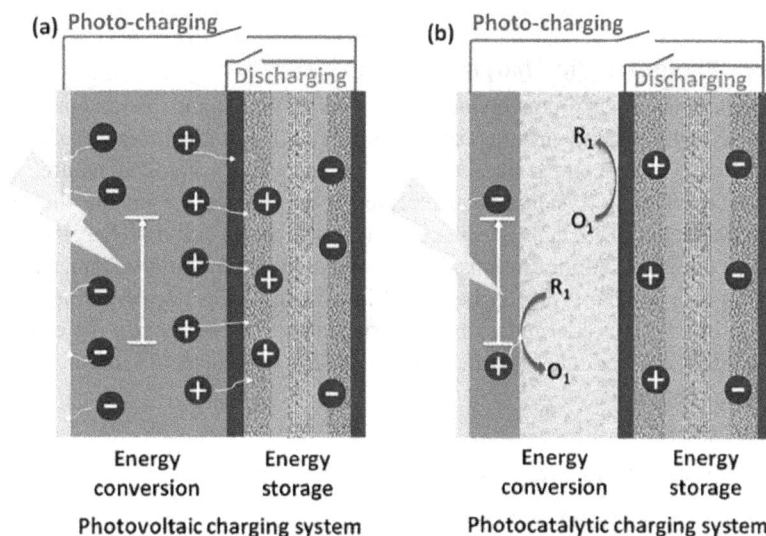

FIGURE 8.3 Two categories of PV-based IECSSs are shown schematically: (a) photovoltaic charging technologies and (b) photocatalytic charging technologies.

8.2.1.1 Photovoltaic Charging System

To achieve practical solar energy storage, recent developments have seen the development of a wide range of integrated systems that combine different PV cell technologies (such as silicon-based solar cells, organic solar cells, and PSCs) with various energy storage solutions (for example, supercapacitors, lithium-ion batteries, and nickel metal hydride batteries). Connecting the energy conversion and storage units using wires is the simplest approach to integrate them (Xu et al., 2015; Yan et al., 2014). For example, a thin amorphous Si PV cell was coupled with a Li-ion battery of the iron phosphate type, for instance, according to Gibson and Kelly. Overall effectiveness of solar energy storage and conversion for the combined system was 14.5% (Gibson and Kelly, 2010). The PV voltage, initially low, was raised to almost 300 V through a DC–DC converter, leading to the charging of the high-voltage NiMH battery pack. This resulted in the creation of an integrated system with high efficiencies in PV -to-battery power storage (Liu et al., 2012). Recent advances in high-performance PSCs have increased opportunities for the creation of solar cells that are both highly efficient (up to 20%) and inexpensive, as well as for the incorporation of solar cells into IECSSs for use in real-world applications (Gurung and Qiao, 2018; Stroyuk, 2018).

The wire was attached with IECSSs that provide several benefits when choosing and assembling individual units. However, the considerable separation between the elements responsible for energy conversion and storage could potentially reduce the overall efficiency of energy storage. One way to get around this problem is to combine the energy conversion and storage components into a single unit, which could also improve space efficiency and boost the system's volume energy density (Wee et al., 2011).

8.2.1.2 Photocatalytic Charging System

Numerous solar energy conversion devices like DSSCs and photocatalytic reactions of solar-hydrogen synthesis have been thoroughly studied (Nazeeruddin et al., 2011). Due to their comparable construction and electrochemical nature, DSSCs and capacitor/battery systems have received a lot of attention during the past ten years (Grätzel et al., 2012). Because of this, most photocatalytic charging systems are based on DSSC configurations. In a photocatalytic charging system, the photo-generated charges can typically be stored in a variety of ways, involving chemical reactions of the electrode materials, electrochemical interactions in the electrolytic solution, double-layer charges processes on the electrode–electrolyte interface, or a combination of these.

8.2.2 BIOENERGY TECHNOLOGIES

Bioenergy is power produced from non-fossilized resources of a biological origin. Raw woody biomass and animal waste currently meet a significant portion of the world's primary energy requirements, with the potential for transformation into various gaseous or liquid fuels.

8.2.2.1 Biogas

Biogas is a concoction of gases produced when organic matter breaks down without oxygen. It can be used to heat or cook food, power internal combustion engines, or be synthesized into synthetic petrol to create superior fuels or chemicals. In order to convert the solid biomass into greenhouse gases (GHGs), it gets heated in without the presence of air, which is subsequently cleaned and filtered to remove undesirable compounds (Uriarte, 2010). Another method for converting organic matter into renewable energy is anaerobic digestion, which is a type of fermentation. Anaerobic biogas consists of 60% methane, akin to landfill gas, with the remaining 40% comprised of carbon dioxide.

8.2.2.2 Biofuels

Examples of liquid biofuels include fuel oil, biogasoline, and pure plant oil, also referred to as straight vegetable oil. The main sources of ethanol production are starch-rich crops like maize, sugar, and corn. Long-chain alkyl esters found in biodiesels are diesel alternatives made from vegetable or animal fats that are typically created by reacting fatty substances with alcohol (Islam et al., 2019). Regular diesel engines can run on biodiesel in its pure form or in any ratio of biodiesel to petroleum diesel; for example, B100 stands for 100% pure biodiesel, B20 for 20% biodiesel mixed with 80% petroleum diesel, and so on (Ackermann, 2005).

8.2.3 Ocean Energy Technology

The vast energy present in ocean waves and tides can be harnessed for practical applications. Additionally, the shift in salinity at river mouths and the temperature difference between cold ocean depths and hot ocean waters can also be faced for the generation of electricity (Adimazoya and Doelle, 2021). The following are the current and future ocean or marine energy technologies.

8.2.3.1 Ocean Thermal Energy Conversion

Ocean Thermal Energy Conversion (OTEC) technology generates electricity by harnessing the temperature contrast between the warmer surface water of a marine lake and the colder waters beneath it (Hamedi and Sadeghzadeh, 2017). It has been found that operative energy can be produced even with a temperature differential of 20°C.

Claude cycle (open), Anderson cycle (closed), and hybrid cycle OTEC plants are the three different types. The Claude cycle uses hot water on the surface from tropical oceans as a heat transfer fluid (HTF). In a vessel that has been partially evacuated, hot water is permitted to expand swiftly and turn into steam. As the generated steam undergoes further expansion, it propels a low-pressure steam turbine. The vapour is condensed into freshwater with no salt at the turbine outflow using cool seawater, which can serve as a source of drinking water for neighbouring settlements. In the Anderson cycle, Freon, propane, or ammonia serves as the active HTF, and the entire system, including the turbine, is fully enclosed or sealed. The fluid is heated through a heat exchanger using hot surface water, and the vapour produced by this process is expanded over the blades of a turbine, which in turn powers a generator. The vapour is fed back to the evaporator after condensing with cool deep ocean water at the turbine exit. In hybrid cycles, heated water from closed-cycle OTECs is injected into an open-cycle OTEC system to increase power output rather than being discharged to the ocean (Islam et al., 2019).

8.2.3.2 Wave Energy

Ocean waves derive their energy from the combination of kinetic and gravitational potential energy, generated when wind moves across the water's surface. Wave energy availability is closely linked to wave height, and during storms, peak wave power can surpass 200 kW/m. Due to their unpredictability and the potential hazards they pose, wave resources are challenging and unsafe for navigation or manipulation (Reddy, 2017).

The energy from ocean waves can be captured by employing a water column in oscillation, which includes an air chamber interfacing with the ocean and a turbine. The oscillation of the water column induces the flow of air into and out of the column, facilitating the collection of wave energy (Islam et al., 2019). To transform wave oscillation motion into linear motion, well compressors with identical airfoil structures edges are counterbalanced by flywheels.

8.2.3.3 Tidal Power

Massive amounts of potential energy are present during tidal movements, and this potential energy can be used to drive a turbogenerator's input or outflow (or both) of seawater (Dye, 2012). A dam is built to separate tidal basins from the seafloor, creating in the water level disparity between the two. At high tide, water surges from the sea into the basin, propelling via a turbine to generate electricity. Conversely, at low tide, water exists of the tidal basin into the sea, causing the turbine to rotate in the different direction and generate power (Islam et al., 2019).

8.3 LIFE CYCLE ANALYSIS

Upon understanding the function of energy conversion and storage system, the next step is to understand their impact to the environment from the creation to the end of life of the devices or systems. A method known as life cycle analysis (LCA) is used to measure the environmental effects of the manufacturing, disposal, and consumption of a product or process. Creating up-to-date environmentally friendly choices can benefit greatly from LCA: (1) products are contrasted according to categories of environmental impact that can be conceptualized by actual environmental harm, (2) it is possible to identify unforeseen environmental trade-offs between effect categories, and (3) a standardized methodology enables life cycle assessments from various studies to be utilized to compare various product options (Guinee, 2002).

One of the methodologies currently available to examine the environmental impacts of construction processes is life cycle assessment analysis. It is a method for assessing how well a thing, activity, or procedure performs from the standpoint of how the environment affects it at each stage. The numerous evaluation processes are carried out by the LCA under ISO 14040's authorization and supervision.

The Society for Environmental Toxicology and Chemistry (SETAC) defines LCA as an evaluation procedure intended to (Anon, n.d.):

- Evaluate the materials and energy used and emitted into the environment to determine environmental issues;
- Identify the environmental impacts of these resources and energy; and
- Find potential environmental development strategies.

For example, LCA for PV devices can be generally summarized as in Tables 8.2–8.4. Though the tables do not provide details breakdown of the LCA information, they present the gist of the findings in terms of advantages and disadvantages. Discussion on LCA of solar cell devices is presented in the subsequent sub-sections.

8.3.1 LIFE CYCLE ANALYSIS OF BATTERIES

An assessment methodology for the entire lifecycle is used to evaluate the battery life cycle studies that were evaluated below.

Specifically, the assessment includes studies that offer comprehensive presentations of process-specific data, particularly those featuring structured process flows or references to them. Figure 8.4 provides an example of the essential flows required to define a unit process. For example, the production of a PbA battery necessitates unit processes for generating lead, acid, battery casings, stakes, barriers, copper, and other components. Additionally, one or more processes are required to assemble these components into a finished product ready for sale (Sullivan and Gaines, 2010). Moreover, diverse unit processes are essential for material manufacturing. To illustrate, mines, beneficiation, ore preparation, casting, and refining represent the required unit operations for lead production. Figure 8.5 illustrates the full life cycle of products (batteries).

TABLE 8.2
The Earliest Photovoltaic Cells

Solar Cells Based on Silicon, or First-Generation PV Cells (–Si)

Due to the superior electrical, chemical, and mechanical qualities, Silicon remains the predominant material utilized in photovoltaic (PV) modules. Adapted from Aberle (2006). Some of the most developed solar technologies rely on this semiconductor.

Type	Efficiency	Advantages	Disadvantages
sc-Si	• The maximum efficiency when compared to other solar technologies was between 25% and 27% in a laboratory (Sørensen, 2017). • The band gap is 1.11–1.15 eV. • The commercial efficiency ranges from 16% to 22%.	The most efficient and pure sort of solar panel is this one.	The manufacturing process (CZ process) requires a lot of resources and energy.
mc-Si	• 15%–18% (Hossam-Eldin et al., 2015) • 1.11 eV serves as the band gap	A good substitute to lower the price of PV modules.	Ineffective compared to sc-Si cells.

TABLE 8.3
PV Cells of the Second Generation

Thin-Film Solar Cells (TFSCs), A Second Generation of Photovoltaic cells

The thickness of film layers can be measured in nanometres (nm) or micrometres. Because of the minimal semiconductor material required for each cell, this method boasts an economical manufacturing cost and low material expenses. In comparison to first-generation PV, TFSC are flexible, lighter, and have less drag thanks to their low thickness. By using sputtering equipment to deposit tiny layers of certain materials on glass, plastic, or stainless-steel substrates, TFSCs are created. A few of these technologies are still in the development stage (Muteri et al., 2020).

Type	Efficiency	Advantages	Disadvantages
a-Si	• 4%–8% for modules commercialized • small cells that are tested in the lab can reach 12% (Hossam-Eldin et al., 2015)	Least expensive	As the layers are thinner and contain less material for absorbing solar radiation compared to single-crystal silicon (sc-Si) and multicrystalline silicon (mc-Si) solar modules, they yield lower electricity production (15%–35%) under sunlight exposure (Ratner, 2004).
CdTe	• 1.45 eV is the band gap value • 21% record efficiency (Ratner, 2004) • 10%–15%	Compared to Si cells, CdTe cells can use a wider wavelength spectrum that is similar to the organic spectrum. Low prices because cadmium is plentiful and produced as a result of the extraction of crucial industrial minerals like zinc.	One of the most dangerous substances is cadmium, and cadmium telluride also has certain hazardous qualities.
CIGS	• 20% (in some circumstances) • The energy band gap is 1.68 electron volts (eV)	The process consumes less energy than the production of solar cells made of crystalline silicon and exhibits excellent heat resistance.	• CIGS solar cells are less effective than Si solar cells and use hazardous chemicals. • Currently, it costs a lot to produce.
GaAs	• In a laboratory, 29% efficiency set a record (Ratner, 2004) • 1.43 eV is the band gap value	Greater effectiveness and thinner than silicon ones.	Cost is high.
CIS	10%–13%	Methodologies use less energy than crystalline technology fabrication, and they have better heat resistance than silicon-based modules (Gul et al., 2016).	Comparatively pricey because of the materials used.

TABLE 8.4
PV Cells of the Third Generation

Third-Generation Solar Cells

The majority of the novel technologies used in third-generation solar panels are still in the research and development stages. Some cells use natural, quasi-organic, or inorganic substrates to produce energy. They also use hybrid systems or novel technological methods built on components at the nanometre and molecular scale (Muteri et al., 2020).

Type	Efficiency	Advantages	Disadvantages
DSSC	Roughly 10% (Muteri et al., 2020)	• Flexibility, lack of pollution, and simplicity of recycling. • Low cost because of the straightforward manufacturing method. • DSSC function even in low light. • Good efficiency also at high temperatures.	• The electrolyte comprises organic solvents with volatility and requires secure sealing due to its susceptibility to freezing at low temperatures, leading to diminished power production and potential physical damage.
QDs	Roughly 1.9%	Simple to produce and prepare.	Low efficiency
PSC	19%–22% (Ratner, 2004)	• Good performance and room for development. • Producing perovskite is less expensive than silicon.	• When exposed to heat, snow, moisture, etc., perovskite degrades quickly. • The existence of lead in perovskite is hotly contested due to toxicity issues.

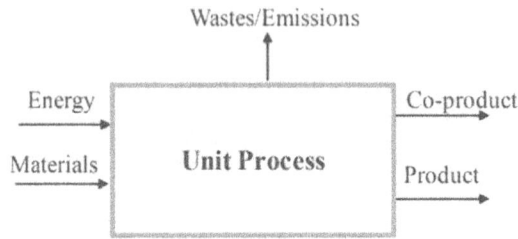

FIGURE 8.4   General unit process. Adapted from Sullivan and Gaines (2010).

FIGURE 8.5   Limitations considered for cradle-to-gate study assessment. Adapted from Sullivan and Gaines (2010).

8.3.1.1   Lead-Acid Batteries

The basic battery materials must be transformed into the materials needed for the battery, which requires a significant amount of energy during battery manufacturing. In actuality, the manufacturing processes involve the production of grids, paste, plates, plastic moulding, and assembly. Lead peroxide and sponge lead are produced through electrolysis processing the lead oxide described first into a paste. The products from paste processing are then placed on lead grids, which are made with energy as well. About 9.2 MJ (30%) of the 31 MJ of energy (Ectg) necessary to produce a kilogram of PbAs battery is used to create the battery (Hittman Associates, 1980).

In his literature, Rantik elucidates the methods required for battery production, covering aspects such as grid creation, lead oxide synthesis, and paste formulation, as well as pasting, drying, curing, and formatting processes. As an illustrative example, he employs an Emnf value of 16.6 MJ/kg (Sullivan and Gaines, 2012).

8.3.1.2   Nickel–Cadmium Batteries

The following manufacturing procedures are necessary to create these batteries (Rantik, 1999). (1) to create sponge nickel, deposit and burn carbonyl Ni powder onto the cathode metal film in a reduction furnace; (2) $Ni(NO_3)_2$ should next be infused into the resultant cathode strip to produce $Ni(OH)_2$; (3) the anode base (nickel fibre grid) is coated with plastic-bonded (PTFE) Cd that has been mixed with a portion of graphite to increase porosity and, consequently, conductivity; (4) create the separators (three thin polymer strips made of nylon, polypropylene, and nylon); (5) the electrodes are charged with surplus electrolyte; (6) arrange separator strips between layers of cathode and anode that are alternately oriented; (7) shape the thermoplastic case; and (8) add the elements to the polypropylene case, such as the electrolyte and seal. For either crystalline or jelly-roll-designed batteries, this set of procedures are used. The Emnf range given by Rydh and Sanden (2005) is 46–63 MJ/kg of battery. They resisted talking about the manufacturing process, though.

8.3.1.3   Lithium-Ion Batteries

These batteries are made using a variety of procedures, which include: (1) manufacture of cathode pastes and cathodes using LiMexOy (Me = Ni, Co, Fe, Mn), binding agents, aluminium sheets, and solvent bought lithium metal oxides; (2) the creation of anodes using strips of copper and graphite pastes; (3) arrangement of separator sheets separating the anodes and cathodes; (4) Electrolyte addition, cell charging, and final assembly come next (Rydh, 2019). Ishihara et al. (2002) in their review found Li-ion Emnf values are around 30 MJ/kg and a high set greater than 100 MJ/kg.

8.3.1.4   Sodium-Sulphur Batteries

These batteries function at elevated temperatures and necessitate insulation to efficiently isolate the high-temperature interior from the surrounding environment, such as through the use of hollow crystal spheres. According to the Hittman estimate, Emnf is responsible for 25% of the overall production energy (Dell'Era et al., 2015).

   In conclusion, life cycle energy values for some of the four battery types covered in this article can be found. PbA and NiCd batteries have values that are quite constant within every generation according to the literature. Na/S battery values are in uncertainty. For Li-ion on the other hand, a wide variety of Emnf values are discovered. To sum up, insufficient quantitative data on the energy and material flow during the construction of various batteries, particularly complex battery systems, raises doubts about the reliability of their respective Emnf values and additional life cycle burdens (Bmnf) (Sullivan and Gaines, 2010).

8.3.2   Life Cycle Analysis of Solar Cells

One layer of light-absorbing materials or several layers of light absorbing can be used to make solar cells. The first scenario involves single junction cells, which are straightforward in their production but not as efficient; in contrast, many configurations (multi-junctions) feature varied charge separation and absorption methods, making them more complicated and effective (Brutschy, 2000).

Cells made from single-crystalline silicon (sc-Si) and multicrystalline silicon (mc-Si), as well as conventional panels with crystalline silicon (c-Si) bases, are included in the first-generation.

The amorphous silicon type (a-Si), the cadmium telluride (CdTe), the cadmium sulphide (CdS), and copper indium gallium selenide (CIGS)/copper indium selenide (CIS) modules based on silicon are among the thin-film solar cells used in the second generation.

The third generation, sometimes referred to as the "next generation," includes novel quasi-silicon-based technologies and innovative devices, including biodynamic/semi-organic photovoltaic panels, PSCs, dye-sensitized solar cells (DSSCs), and quantum dot (QD) cells.

8.3.2.1   Silicon Solar Cells

A metalloid called silicon was found in 1824 (Bagher et al., 2015). Most glasses, high-tech devices, semiconductor components for computers, and auto parts are all made from this metalloid, which is the most abundant semiconductor in the world. PV modules made of crystalline silicon are created using a number of methods.

In an arc furnace, silicon dioxide, also known as SiO_2 or quartz sand silica, undergoes conversion into metallurgical-grade silicon (MG-Si) (Pappas, 2016). Additionally, silicon must undergo refinement to achieve silicon of solar-grade purity (>99.99%) can be produced through diverse methods, including the Siemens process and its modified versions. Operating at high temperatures (usually ranging between 1,100°C and 1,200°C) in the reaction chamber, this process consumes a substantial amount of energy to produce highly pure silicon (Peng et al., 2013).

With Japan's solar irradiance measuring 1,427 kWh/m²/year, Kato et al. conducted an LCA on a 3-kWp home PV system installed on a rooftop utilizing off-grid monocrystalline silicon (mono-Si), multicrystalline silicon (multi-Si), and amorphous silicon (a-Si) (Wong et al., 2016). Based on the findings, mono-Si modules demonstrate an Energy Payback Time (EPBT) of 11.8 years and a GHG emissions rate of 60.1 g CO_2-eq/kWh for an annual production of 10 MW/year. In comparison, multi-Si modules and a-Si modules, both with an equivalent annual cell production of 10 MW/year, were projected to have shorter EPBTs of 2.4 years and 2.1 years, respectively (Figure 8.6).

Kannan et al. explored the EPBT and GHG emissions of a 2.7 kWp mono-Si PV system under Singapore's solar irradiation of 1,635 kWh/m²/year. The EPBT and GHG emissions determined through LCA were identified as 6.74 years and 217 g CO_2-eq/kWh, respectively (Kato et al., 2017). Moreover, in a comparative LCA conducted by Wei Luo et al. (2018), three roof-integrated multicrystalline PV systems—AL-BSF (aluminium back surface field), PERC (passivated emitter and rear cell), and PERC solar cells with the frameless double-glass module structure—were investigated. The results revealed EPBTs of 1.11, 1.08, and 1.01 years, along with associated GHG emissions rates of 30.2, 29.2, and 20.9 g CO_2-eq/kWh for the respective PV systems mentioned (Kannan et al., 2006). Frameless double-glass PV module designs have been found to significantly lower EPBT and perform better in terms of GHG emissions.

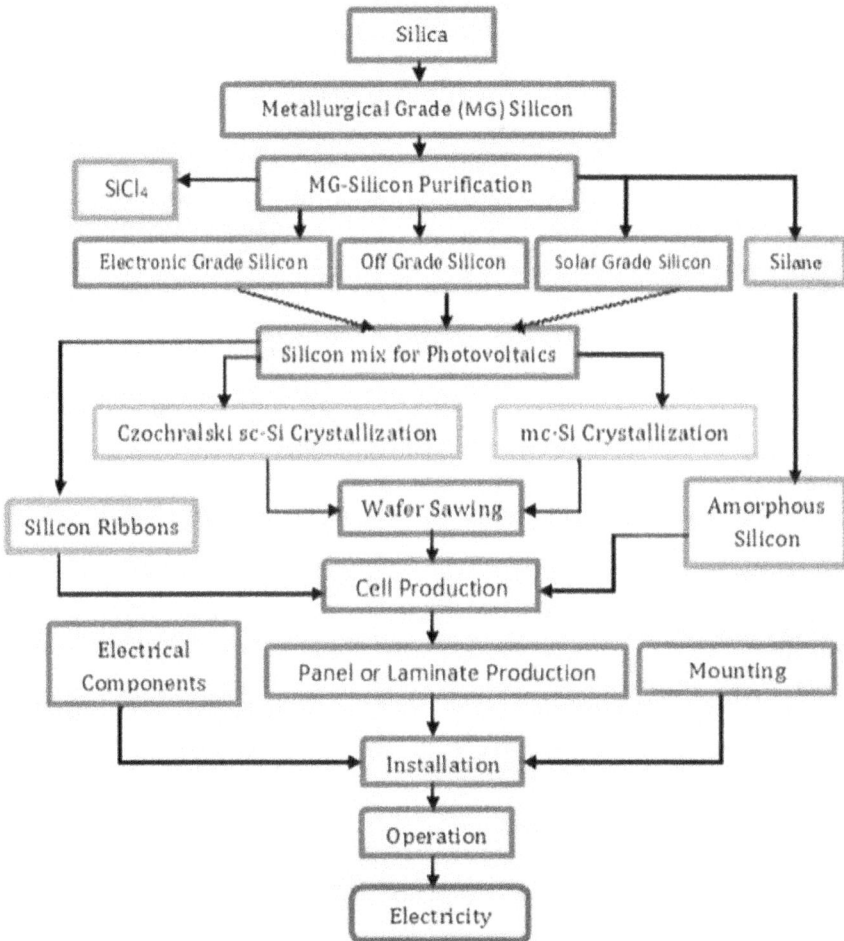

FIGURE 8.6 Stages of silicon-based PV components' life cycles. Adapted from Peng et al. (2013).

8.3.2.2 Thin-Film Solar Cells

The first LCA on these kinds of PVs was carried out in 2005 using 9% efficient CdTe modules. For the first solar facility in the US, LCI is carried out while accounting for the processes from raw materials including the BOS elements' removal to PV installation. The GHG emissions are 23.6 gCO_2-eq/kW h, and the EPBT is 1.2 years. Compared to silicon panels, this type has a smaller ecological footprint (Luo et al., 2018). Fthenakis outlined the phases of the life cycle of CdTe PV, specifically during thin-film production utilizing the cadmium telluride (CdTe) method (Nayeripour and Kheshti, 2011).

Most of the tellurium (Te) was extracted from the residue, known as slime, remaining after the electrolysis-based refining of copper. This residue contained copper

(Cu), selenium (Se), and other metals. The extraction of cadmium (Cd) from zinc ores (80%) and tellurium from copper ores was initially performed (Fthenakis et al., 2008; Alami, 2023). Following electrolytic purification, evaporating or distillation under vacuum were used to create >99.99% pure Cd and Te, which was subsequently used to synthesize CdTe. Following the initial application of a transparent conducting oxide (TCO) layer, a thin layer of CdS and subsequently a layer of CdTe were vapour-deposited onto a glass substrate. Subsequently, a heat treatment was applied after spraying $CdCl_2$. The CdTe solar cell was finalized through the deposition of a metal layer using vacuum arc-based methods to form the rear contact (Singh et al., 2013). Before becoming operational and generating electricity, the installation of the PV module, inverter, electrical components, and support systems took place. The flowchart for the manufacture of thin-film PV modules is shown in Figure 8.7 (Pappas, 2016).

To assess a-Si and multi-Si at various annual production rates, Kato et al. (2001) conducted an LCA on CdS/CdTe PV modules installed on residential rooftops to estimate primary energy consumption, CO_2, and EPBT emissions. According to the study, CdS/CdTe PV modules had EPBTs that were lower than those of a-Si and multi-Si, measuring 1.7 years at 10 MW/year, 1.4 years at 30 MW/year, and 1.1 years at 100 MW/year. The CdS/CdTe PV modules have life cycle GHG emissions ranging from 8.9 to 14.0 g CO_2-eq/kWh for annual production between 10 and 100 MW/year. In the Gobi Desert, Ito et al. (2008) performed a comparative LCA research for a 100-MW VLS-PV system employing four alternative PV module types: multi-Si, a-Si, CdTe, and CIS at various module efficiencies. In comparison to CdTe and a-Si modules, which had EPBTs of 1.9 years and 2.5 years, respectively, the scientists found that CIS modules have shorter EPBTs of 1.6 years. These modules' GHG emissions ranged from 9 to 16 g CO_2-eq/kWh.

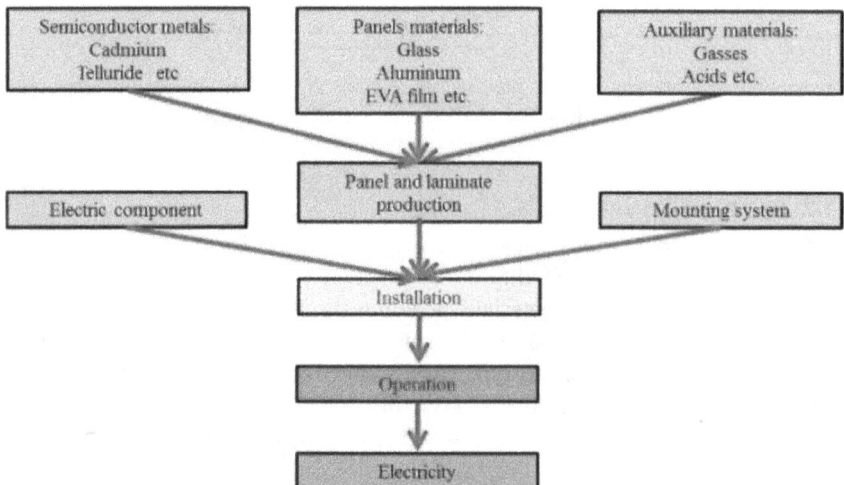

FIGURE 8.7 Stages of thin-film PV components' life cycles. Adapted from Pappas (2016).

In a 33 kW rooftop installation utilizing KC120 (multi-Si) and PV136 (a-Si) modules in Ann Arbor, United States, with solar irradiation levels of 1,359 kWh/m^2/year, Pacca et al. (2007) analysed three LCA parameters, namely, Net Energy Ratio (NER), EPBT, and GHG emissions. As per the study, the EPBT for a-Si is lower than that of multi-Si (3.2 years vs. 7.5 years). Yet, with potential improvements in conversion efficiency in the future, the anticipated EPBT for a-Si and multi-Si could be 1.6 and 5.7 years, respectively. The GHG emission rate follows a similar trend, with the a-Si GHG emission rate at 34.3 g CO_2-eq/kWh, considerably less than the multi-Si GHG emission rate of 72.4 g CO_2-eq/kWh (Ludin et al., 2018).

8.3.2.3 Dye-Sensitized Solar Cell (DSSC)

Due to their potential for having a lesser economic and environmental impact than standard PV technologies, DSSCs have some differentiation advantages. Initially, TCO coatings are applied to glass substrates, nanoporous titanium dioxide (TiO_2), which serves as a dye transporter—typically fluorine-doped tin oxide (FTO)—in the production of DSSCs (Luque and Hegedus, 2003). After the deposition, the photoelectrode undergoes heating in a furnace to sinter the TiO_2, and the TiO_2 surface is subsequently coloured using dye sensitizer molecules, such as natural dye (Gong et al., 2012). The electrolyte solution, occupying the dye, contains I$^-$/I3$^-$ redox ions along with minimal-viscosity solvents that modify ion conductivity, influencing the cell's performance (Luque and Hegedus, 2003). Platinum-coated counter-electrodes on other indium tin oxide glass sheets facilitate electron collection, completing the DSSC cell (Gong et al., 2012). Following this, DSSC cells are transformed into modules, subsequently mounted, and wired into the grid. The system boundaries for DSSC LCA experiments are depicted in Figure 8.8.

An environmental evaluation at laboratory scale was done on the production of DSSC PV in Southern Europe by Parisi et al. (2014). This analysis made the following assumptions about the DSSC module: an efficiency of 8%, a lifetime of 20 years, a performance ratio (PR) of 0.75, and the usage of the European Union for Coordination of Transmission Electricity as the module's energy source. The NER for the DSSC module stood at 12.67, with corresponding EPBT and CO_2 emissions of 1.58 years and 22.38 g CO_2-eq/kWh, respectively, according to the authors' findings. The LCA of the DSSC study was then compared to other chemical and biological thin-film technologies (Parisi et al., 2014). In a subsequent LCA on DSSCs from earlier investigations, Parisi et al. (2014) compared the DSSC with alternative thin-film technologies, including polymers, a-Si, CdTe, and CIS (de Wild-Scholten and Velkamp, 2007). GHG emissions rate ranged from 277 to 365 MJ/m^2, 0.6 to 1.8 years, and 9.8 to 120.0 g CO_2-eq/kWh, respectively. Notably, there was a strong correlation between DSSC GHG emissions and module functional lifetime (de Wild-Scholten and Velkamp, 2007). Longer lives and higher conversion efficiencies could lead to the DSSC's optimum performance.

8.3.2.4 Perovskite Solar Cell (PSC)

PSCs have gained considerable attention in recent years as one of the newly developed PV technologies that have excellent productivity with minimal cell production costs. In earlier LCA analyses of PSCs, Figure 8.9 depicts the system boundaries

Input   System Boundary   Output

Raw Materials:
a) FTO glass substrate (x2)
b) Semiconductor TiO2
c) Platinum
d) Metallization silver paste
e) Dye
f) Electrolyte

Raw Materials   Emissions

DSSC Cell Production Process:
a) Glass substrates coating
b) Substrate etching
c) Deposition of active layer/ TiO2
    screen printing
d) Sintering
e) Staining
f) Substrate matching
g) Electrolyte filling

Energy   Waste
Consumption   treatment/
                                                                   Disposal
DSSC Module Manufacturing:
a) Completing the cell
b) Panel manufacturing
c) Testing
d) Final Q&A
e) Packaging

DSSC PV System installation:
1) Support structures
2) Mounted PV systems
3) PV modules cables, and power
    conditioning equipment are
    integrated (BOS)

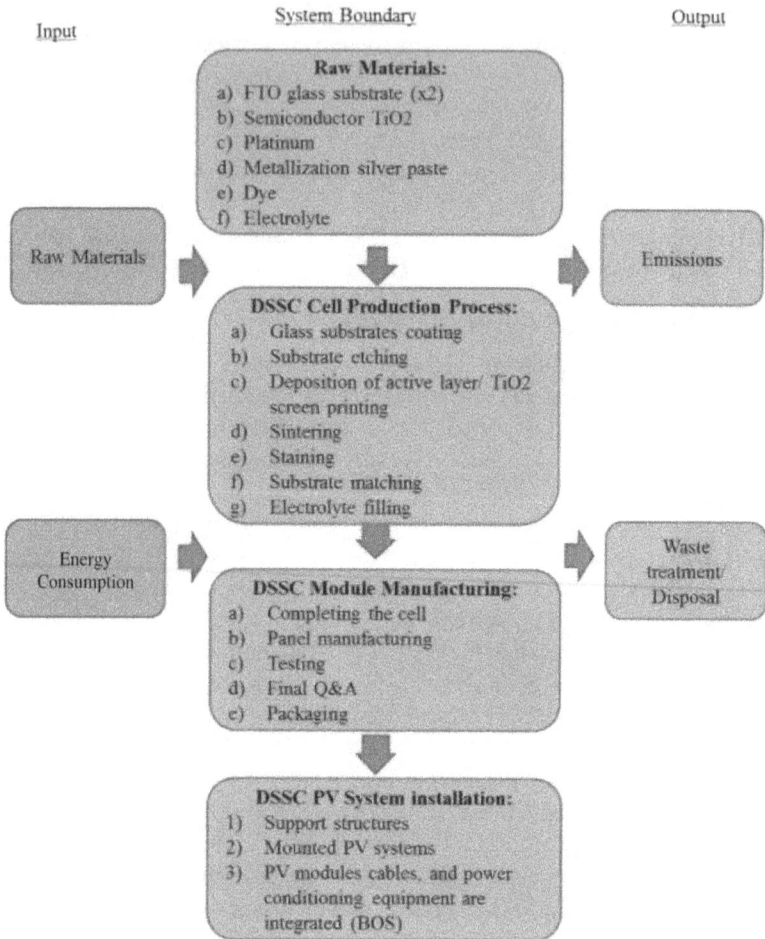

FIGURE 8.8 LCA from cradle to gate for the DSSC procedure. Adapted from Ludin et al. (2018).

taking into account a cradle-to-grave approach, encompassing the production of the individual modules, their use, and their disposal (Gong et al., 2015).

The initial phase commences with the extraction of raw materials required for manufacturing components essential in the subsequent stage. These materials include FTO glass, BL-TiO$_2$ ink, PbI$_2$, gold, and PET production. There are multiple phases involved in making a perovskite PV module, beginning with FTO glass manufacturing, substrates, blocking TiO$_2$ layer formation, followed by the process involves depositing an electron transport layer, a perovskite layer, and a hole transport layer on the substrate, followed by the thermal evaporation of gold (Gong and Sumathi, 2015). In the usage phase, the perovskite modules were used to produce electricity after being put together. In the disposal stage, the waste modules were finally dumped in the ground.

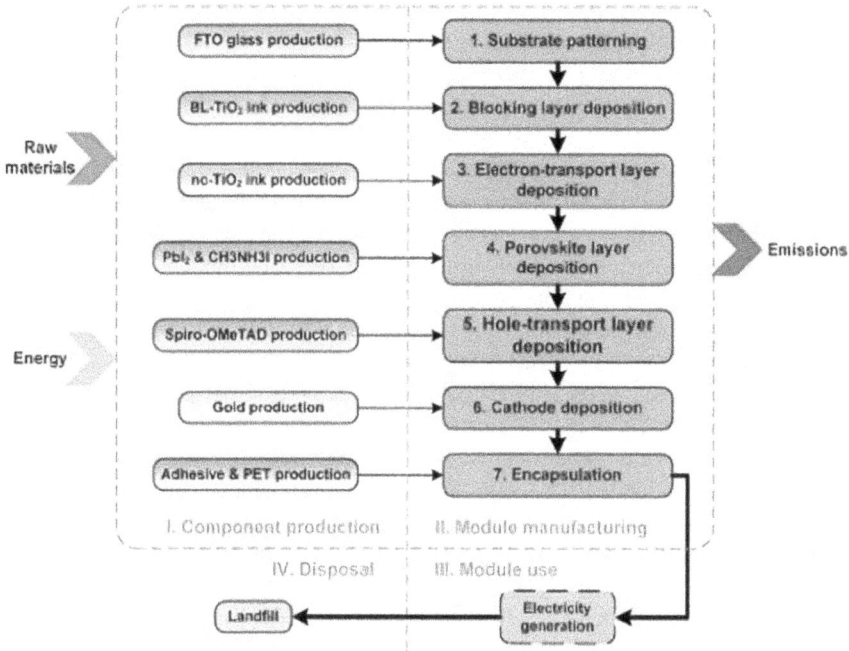

FIGURE 8.9 TiO_2 system boundary in perovskite solar module production. Adapted from Ludin et al. (2018).

Assuming a 15% module efficiency, a PR of 0.75, a 15-year lifespan, and 65% active area under solar irradiation of 1,700 kWh/m^2/year, Celik et al. assessed the environmental impact of three PSCs structures: void, solution, and hole transfer layer-free devices (Celik et al., 2016). The findings indicate that, in comparison to advancements in crystalline and thin-layer technologies, the PSCs exhibited EPBT and GHG emissions ranging from 1.05 to 1.54 years and 100–150 g CO_2-eq/kWh, respectively.

The EPBT spanned from 0.2 to 5.4 years, while the GHG emissions rate varied from 56.65 to 497.2 g CO_2-eq/kWh. Notably, PSCs demonstrated significantly lower energy consumption compared to silicon and lightweight innovations, showcasing competitive environmental advantages and an EPBT that paves the way for future production (Ludin et al., 2018).

8.4   RECYCLING AS SUSTAINABLE USAGE OF MATERIALS

The fast growth of the developer's material palette over the past few decades has hampered the recycling of items in the "occasionally abandoned" or "owned by someone else" categories (Zhang et al., 2010; Rozendal et al., 2006). As a result, many goods are now more durable and efficient than before. Recycling has become much more complicated and difficult to accomplish, which was not the desired outcome. In recent years, several reviews of metal recycling have been published. They explore important topics like reuse and recycling of technological devices, financial constraints, and improvement techniques (Ayres et al., 2006). Some unanswered queries are as follows: What's happening, and what are the recent developments? What are the boundaries? Can there be an integrated materials economy?

How successful is the globe recycling the wide range of components found in contemporary goods? Recycling at the end-of-life rate and recycled materials are the two indicators that best address this issue. Recycled content specifies the percentage of scrap used in the manufacturing of metal, which is crucial to understanding the size of the reclaimed supplier. Nevertheless, this indicator has two drawbacks. Initially given that the lifespans of metal-containing items frequently extend several decades and that metal use is expanding quickly, recycled metal streams will only be able to satisfy a small percentage of demand for a very long period (Logan and Rabaey, 2012). Second, it does not differentiate between new and old (post-consumer) scrap as an input material, leaving it open to artificially increased rates based only on pre-consumer sources (fabricators may be given incentives to increase their discard output to meet supplementary consumer demand, which makes recycled content a justification for shortcomings in fabrication and production). Instead, what recycled content means to promote is the volume of used materials that meet the criteria for regeneration.

Some devices like red phosphors, high-strength magnets, thin-film solar cells, and computer chips, are increasingly used in small quantities for very precise technological applications. Recovery in those applications, which frequently involve finely blended "specialty metals," can be so difficult mechanically and financially that attempts are rarely performed.

8.4.1 ACQUISITION, PRELIMINARY PROCESSING, AND FINISHING OFF

Depending on cost, administrative tasks, and other considerations, acquisition rates for various waste streams might be very varied. Contrarily, despite regulatory efforts, acquisition rates for waste electrical and electronic equipment are frequently quite low. The pre-processing process for blog-consumer metal begins with collecting and includes continuous screening (like manual labour, conductive, or visual methods) and both chemical and physical separation (Cheng et al., 2009).

Scale-related issues are crucial. Production of virgin materials often takes place on a massive scale and is supported by historically cheap energy rates. Recycling, on the other hand, is frequently more regional, difficult to implement, and small scale. The financial benefits in such a situation are frequently insufficient to support the acquisition of contemporary the "impression and kind" technology, and a great deal of otherwise recoverable material is wasted (Logan and Rabaey, 2012).

8.4.2 LEVERAGING THE RECYCLING PROCESS

While segregation and organizing efficiency are tied to recyclable technological advances, acquisition efficiencies are connected to societal and governmental variables. The situation is unfavourable from a materials standpoint that, due to scale and budgetary constraints, more sophisticated technologies (like laser beam, near-infrared, or x-ray sorting) are typically only applied to specific recycling streams. Examples of more basic technologies include splitting, compression, and magnetic sorting. The design of products frequently makes it difficult to disassemble and liberate materials (for example, constructed magnetic strips in desktops) (Nam and Logan, 2011).

Though there are significant examples of cutting-edge repurposing innovations, many of which are still in the demonstration stage, much more focus needs to be placed on transforming and improving existing generic systems if efficiency ratings are to increase from their current levels (Dahmus and Gutowski, 2007). Such technological advancement might coincide with a globalization of labour, as is typical in production operations. For electronics, the finest-of-both-worlds strategy suggests utilizing the cheap labour costs in poor nations for manual disintegration and the high productivity of specialized refineries, which are normally found in industrialized nations for halt-processing (Cheng et al., 2009).

8.4.3 Future Recycling Concerns to Be Addressed

The activity that has the greatest chance to promote the recycling of metal is collecting, despite the fact that it initially appears insignificant. It is mostly a matter of behavioural norms and encouragement, as well as programmes like the requirement of recycling deposits on products for consumption, to collect discards with high effectiveness and with due care (to avoid mixing that would impede subsequent treatment). A further obstacle to the collection and rehabilitation of many elements is the worldwide commerce in used goods, which sends complicated products to nations with insufficient recycling infrastructure (Speers and Reguera, 2012; Stoytcheva, 2011).

The next endeavour is to involve upcoming manufacturers in selecting the right combinations of materials with reusing in mind after taking care of the collection. The current tendency of more material mixing can only be reversed by designers, yet the contemporary designs are actually harder to recycle than they were a few generations prior. The next endeavour is to involve upcoming manufacturers in selecting the right combinations of materials with reusing in mind after taking care of the collection. The current tendency of more material mixing can only be reversed by designers, yet the contemporary designs are actually harder to recycle than they were a few generations prior (Steinbusch et al., 2009; Heidrich et al., 2011).

8.5 NEW OPTIONS OF SUSTAINABLE USAGE IN ENERGY CONVERSION

As discussed briefly above, solar cell devices are the straightforward option of sustainable energy conversion. Other new options include fuel cells and rechargeable batteries.

8.5.1 Fuel Cells: The Conversion of Hydrogen

For the conversion of hydrogen, various technologies have already been established. For instance, hydrogen can be utilized in both fuel cells and engines. Fuel cells are electrical devices that convert chemical energy, typically in the form of hydrogen, into electricity. Engines, on the other hand, can combust hydrogen similarly to the way they burn gasoline or natural gas (Sherif et al., 2005). In a hydrogen fuel cell, the elements hydrogen and oxygen mix electrochemically (reverse electrolysis) without burning, creating electrically charged energy. The primary obstacles to overcome are the high production costs brought on by the pricey components necessary to make electrodes (bipolar plates), electrolytes, membranes, and catalysts (particularly given the price of platinum).

The automotive sector has historically been interested in PEMFC fuel cells for zero-emission cars, as well as the power production industries for both residential and electronic uses. They are intended to be efficient and clean power sources, but their commercial success is hindered by two key factors: the high cost of Pt/support electrodes and the membrane's poor performance, especially at elevated temperatures (Serrano et al., 2009).

By greatly reducing the quantity and endurance of Pt catalyst required for the application, it is possible to significantly lower the overall expense of the electrodes. Additionally, there is a clear connection between this cost and the support's impact on catalyst performance. For polymer electrolyte membrane fuel cells (PEMFC), carbon nanotube–based electrodes have been created by Wang et al. (2004). They discovered that MWNT can be employed as the support for the catalyst rather than carbon powder, allowing for a 2/3 reduction in Pt usage when compared to conventional PEMFC fuel cells.

Other methods rely on creating metal-based nanostructured catalyst employing a variety of later removed self-organized templates as configuration-directing elements. In the context of serving as anodes for Direct Methanol Fuel Cells (DMFC), Choi et al. employed a template based on crystalline structures with nanosized holes. Within these holes, they created Pt-Ru nanoscale systems (Hossain, 2011).

The application of nanotechnology hydrophilic inorganic substances in electrolytes and an inorganic lithium salt as additives to improve the hydrogen ion conductivity of a membrane at high temperatures is one of the contributions of nanotechnology (the inorganic materials present high affinity to water and the salts work in the right temperature range). Other methods rely on the use of capillary fuel cells rather than flat cell membranes, which are significantly less expensive than the latter. As a result, their more compact construction results in a three to six times higher power density and enables the preservation of the electrodes' bipolar plates (Serrano et al., 2009).

8.5.2 BATTERY RECHARGEABLES

Renewable lithium batteries are the current subject of most ongoing research in this area. The Li-ion chemistry results in a rise of 100%–150% in the capacity to store of energy for each unit of mass and volume when compared with hydrophilic batteries. Despite this, there are significant drawbacks, including low power and energy density, a big shift in volume on consequences, security, and expense. The drawbacks can be (or are being) eliminated through the use of nanomaterials in the production of lithium-ion batteries (Serrano et al., 2009). Silicon, alumina, or chromium nanoparticles can boost the conductivity of non-aqueous electrolytic solutions by a factor of six. Solid polymer electrolytes (SPE) and solid-state electrolytes have received the majority of attention. The PEO-based SPE has drawn the greatest attention since PEO is secure, environmentally friendly, and produces flexible films. However, polymers commonly exhibit low conductivities at room temperature, and depending on the SPE designs, they may have insufficient interfacial mobility and stable mechanical properties. In addressing these challenges, the utilization of nanocomposite polymer electrolytes has the potential to contribute to the development of highly efficient, secure, and environmentally friendly batteries (Agrawal and Pandey, 2008).

Graphite granules and nanotubes made of carbon were used in place of $LiC6$ electrodes, which had a capacity to store of 340 mAh/g, in certain efficient methods for nanostructuring the anode for lithium-ion rechargeable batteries. The lithium deposition as well as certain safety issues were avoided by replacing these with nanosized metallic oxides, primarily in the form of nanotubes and nanowires, such as titanium, aluminium, vanadium, cobalt, tin, and silicon oxides. This has allowed for an amount of storage that is comparable to that of conventional graphite electrodes (Nam et al., 2006; Chan et al., 2007). A study by Chan et al. (2007) achieved a charge capacity that was around 3,000 mAh/g for 10 cycles following the first one, shedding roughly one third of capacity in the first cycle, by synthesizing silica nanowires that are further fixed to the substrate functioning as current collector.

Additionally, $LiFePO_4$, $LiMn_2O_4$, and vanadium oxide (V_2O_5) have been successfully used to examine the nanostructuration of the cathode. Manganese oxides are a good option for both the environment and the economy. $LiMn_2O_4$ performs the best when used in large quantities. For this, many nanostructures have been examined. Since nanoparticles need to be coated to prevent sinterization and increase conductivity, they are less desired (Serrano et al., 2009).

8.6 CONCLUSION

We must continue to create, transform, and use energy responsibly while minimizing our negative environmental effects because the world's population is growing and has an increasing need for energy. The development of novel multifunctional materials from the vast and interdisciplinary discipline known as nanotechnology today is essential for overcoming least of the technological constraints of the many non-renewable energy options. PV solar cells' efficiency is rising as their production and electricity production costs are falling at a pace never seen before due to improved nanomaterials. Numerous technological advances are required to transition from an energy system based on carbon to one that is more sustainable, including those in the production of energy, shipping, alteration, storage, and final use. We face considerable scientific and engineering obstacles across all these processes. The materials we currently have are frequently unable to offer an alternative at the level of effectiveness necessary at a workable cost.

ACKNOWLEDGEMENT

The research was supported by the Ministry of Higher Education (MoHE), through the Fundamental Research Grant Scheme (FRGS/1/2022/TK08/UTAR/02/7).

REFERENCES

Aberle, A.G. (2006). Fabrication and characterisation of crystalline silicon thin-film materials for solar cells. *Thin Solid Films* 511–512, 26–34.

Ackermann, T. (2005). *Wind Power in Power Systems*, John Wiley and Sons, Hoboken, pp. 58–110.

Adeney, K. (2002). *Ethnic Conflict and Civic Life: Hindus and Muslims in India, Ashutosh Varshney*, Yale University Press, New Haven, pp. 157–159.

Adimazoya, T.N., and Doelle, M. (2021). Regulating Wave, Tidal and Ocean Thermal EnergyIn: *Elgar Encyclopedia of Environmental Law*, ed. Michael F., pp. 546–557.

Agrawal, R.C., and Pandey, G.P. (2008). Solid polymer electrolytes: Materials designing and all-solid-state battery applications: An overview. *Journal of Physics D: Applied Physics* 41(22), 223001.

Alami, A.H. (2023). *PV Technology and Manufacturing*. Springer Nature, New York.

Anon, Trove. Trove. https://trove.nla.gov.au/work/15414012 (accessed: Oct 2023)

Ayres, R.U., Gleich, A., and Von Gößling-Reisemann, S. (2007). *Sustainable Metals Management : Securing our Future-Steps Towards a Closed Loop Economy, Eco-Efficiency in Industry and Science, XVI*, 610 p.

Bagher, A. M., Vahid, M. M. A., and Mohsen, M. (2015). Types of solar cells and application. *American Journal of optics and Photonics*, 3(5), 94–113.

Basosi, R., and Parisi, M.L. (2015). Environmental life cycle analysis of nonconventional thin-film photovoltaics: The case of dye-sensitized solar devices. In: *Energy Security and Development: The Global Context and Indian Perspectives* ed. Reddy, B., Ulgiati, S.. Springer, India, pp. 193–210.

Brutschy, B. (2000). Structure of microsolvated benzene derivatives and the role of aromatic substituents. *Chemical Reviews* 100, 3891–3920.

Celik, I., Song, Z., Cimaroli, A.J., Yan, Y., Heben, M.J., and Apul, D. (2016). Life cycle assessment (LCA) of perovskite PV cells projected from lab to fab. *Solar Energy Materials and Solar Cells* 156, 157–169.

Chan, C.K., Peng, H., Twesten, R.D., Jarausch, K., Zhang, X.F., and Cui, Y. (2007). Fast, completely reversible Li insertion in vanadium pentoxide nanoribbons. *Nano Letters* 7, 490–495.

Chan, C.K., Peng, H., Liu, G., McIlwrath, K., Zhang, X.F., Huggins, R.A., and Cui, Y. (2008). High-performance lithium battery anodes using silicon nanowires. *Nature Nanotechnology* 3, 31–35.

Cheng, S., Xing, D., Call, D.F., and Logan, B.E. (2009). Direct biological conversion of electrical current into methane by electromethanogenesis. *Environmental Science and Technology* 43, 3953–3958.

Costanza, R., D'Arge, R., De Groot, R., Farber, S., Grasso, M., Hannon, B., Limburg, K., Naeem, S., O'Neill, R.V., Paruelo, J., et al. (1997). The value of the world's ecosystem services and natural capital. *Nature* 387, 253–260.

Dahmus, J.B., and Gutowski, T.G. (2007). What gets recycled: An information theory based model for product recycling. *Environmental Science and Technology* 41, 7543–7550.

de Wild-Scholten, M.J., and Veltkamp, A.C. (2007). Environmental life cycle analysis of dye sensitized solar devices: Status and outlook. In: *Presented at: 22nd European Photovoltaic Solar Energy Conference and Exhibition, Milan, Italy* vol. 3, pp. 1–5.

Dell'Era, A., Bocci, E., Sbordone, D., Villarini, M., Boeri, F., Di Carlo, A., and Zuccari, F. (2015). Experimental tests to recover the photovoltaic power by battery system. *Energy Procedia*, 81,. 548–557.

Dye, S.T. (2012). Geoneutrinos and the radioactive power of the Earth. *Reviews of Geophysics* 50, 19.

Fthenakis, V.M., Hyung, C.K., and Alsema, E. (2008). Emissions from photovoltaic life cycles. *Environmental Science and Technology* 42, 2168–2174.

Gibson, T.L., and Kelly, N.A. (2010). Solar photovoltaic charging of lithium-ion batteries. *Journal of Power Sources* 195, 3928–3932.

Graedel, T.E., Allwood, J., Birat, J.P., Buchert, M., Hagelüken, C., Reck, B.K., Sibley, S.F., and Sonnemann, G. (2011). What do we know about metal recycling rates? *Journal of Industrial Ecology* 15, 355–366.

Graetzel, M., Janssen, R.A.J., Mitzi, D.B., and Sargent, E.H. (2012). Materials interface engineering for solution-processed photovoltaics. *Nature* 488, 304–312.

Gong, J., Liang, J., and Sumathy, K. (2012). Review on dye-sensitized solar cells (DSSCs): Fundamental concepts and novel materials. *Renewable and Sustainable Energy Reviews* 16, 5848–5860.

Gong, J., Darling, S.B., and You, F. (2015). Perovskite photovoltaics: Life-cycle assessment of energy and environmental impacts. *Energy and Environmental Science* 8, 1953–1968.

Green, M.L., Espinal, L., Traversa, E., and Amis, E.J. (2012). Materials for sustainable development. *MRS Bulletin* 37, 303–308.

Guinee, J.B. (2002). Handbook on life cycle assessment operational guide to the ISO standards. *The International Journal of Life Cycle Assessment* 7, 311-313.

Gurung, A., and Qiao, Q. (2018). Solar charging batteries: Advances, challenges, and opportunities. *Joule* 2, 1217–1230.

Gul, M., Kotak, Y., and Muneer, T. (2016). Review on recent trend of solar photovoltaic technology. *Energy Exploration and Exploitation* 34, 485–526.

Hamedi, A.S., and Sadeghzadeh, S. (2017). Conceptual design of a 5 MW OTEC power plant in the Oman Sea. *Journal of Marine Engineering and Technology* 16, 94–102.

Heidrich, E.S., Curtis, T.P., and Dolfing, J. (2011). Determination of the internal chemical energy of wastewater. *Environmental Science and Technology* 45, 827–832.

Hittman Associates. (1980). Life cycle energy analysis of electric vehicle storage batteries. H-1008/001-80-964, submitted to the U.S. Department of Energy, contract number DE-AC02-79ET25420.A000

Hossam-Eldin, A. & Refaey, M., and Farghly, A. (2015). A review on photovoltaic solar energy technology and its efficiency. In: *Proceedings of the Conference: 17th International Middle-East Power System Conference (MEPCON'15)*, Mansoura, Egypt, 15–17 December 2015.

Hossain, M.M. (2012). *Heat and Mass Transfer: Modeling and Simulation*, InTech, Rijeka.

Ishihara, K., and Kihira, N. (2002). Environmental burdens of large lithium-ion batteries developed in a Japanese national project. *Power Industry*, Japan 11.

Islam, M.M., Hasanuzzaman, M., Pandey, A.K., and Rahim, N.A. (2019). Modern energy conversion technologies. In: *Energy for Sustainable Development: Demand, Supply, Conversion and Management*, ed. Hasanuzzaman, MD., and Rahim, N.A., Elsevier, Amsterdam, pp. 19–39.

Ito, M., Kato, K., Komoto, K., Kichimi, T., and Kurokawa, K. (2008). A comparative study on cost and life-cycle analysis for 100 MW very large-scale PV (VLS-PV) systems in deserts using m-Si, a-Si, CdTe, and CIS modules. *Progress in Photovoltaics: Research and Applications* 16, 17–30.

Kato, K., Murata, A., and Sakuta, K. (1998). Energy pay-back time and life-cycle CO_2 emission of residential PV power system with silicon PV module. *Progress in Photovoltaics: Research and Applications* 6, 105–115.

Kato, K., Hibino, T., Komoto, K., Ihara, S., Yamamoto, S., and Fujihara, H. (2001). Life-cycle analysis on thin-film CdS/CdTe PV modules. *Solar Energy Materials and Solar Cells* 67, 279–287.

Kannan, R., Leong, K.C., Osman, R., Ho, H.K., and Tso, C.P. (2006). Life cycle assessment study of solar PV systems: An example of a 2.7 kWp distributed solar PV system in Singapore. *Solar Energy* 80, 555–563.

Landi, D., Marconi, M., and Pietroni, G. (2022). Comparative life cycle assessment of two different battery technologies: Lithium iron phosphate and sodium-sulfur. *Procedia CIRP*, 105, 482–488.

Lawton, J.H. (1998). *Nature's Services: Societal Dependence on Natural Ecosystems*, Island Press, Washington, D.C.

LeSar, R., Chen, K., and Apelian, D. (2012). Teaching sustainable development in materials science and engineering. *MRS Bulletin* 37, 449–454.

Liu, P., Yang, H.X., Ai, X.P., Li, G.R., and Gao, X.P. (2012). A solar rechargeable battery based on polymeric charge storage electrodes. *Electrochemistry Communications* 16, 69–72.

Logan, B.E., and Rabaey, K. (2012). Conversion of wastes into bioelectricity and chemicals by using microbial electrochemical technologies. *Science* 337, 686–690.

Ludin, N.A., Mustafa, N.I., Hanafiah, M.M., Ibrahim, M.A., Asri Mat Teridi, M., Sepeai, S., Zaharim, A., and Sopian, K. (2018). Prospects of life cycle assessment of renewable energy from solar photovoltaic technologies: A review. *Renewable and Sustainable Energy Reviews* 96, 11–28.

Luo, B., Ye, D., and Wang, L. (2017). Recent progress on integrated energy conversion and storage systems. *Advanced Science* 4, 1700104.

Luo, W., Khoo, Y.S., Kumar, A., Low, J.S.C., Li, Y., Tan, Y.S., Wang, Y., Aberle, A.G., and Ramakrishna, S. (2018). A comparative life-cycle assessment of photovoltaic electricity generation in Singapore by multicrystalline silicon technologies. *Solar Energy Materials and Solar Cells* 174, 157–162.

Luque, A. and Hegedus, S. (2003). *Handbook of Photovoltaic Science and Engineering*. Wiley, New York.

Muehlebach, A. (2001). "Making place" at the United Nations: Indigenous cultural politics at the U.N. *Working Group on Indigenous Populations. Cultural Anthropology: Journal of the Society for Cultural Anthropology* 16, 415–448.

Muteri, V., Cellura, M., Curto, D., Franzitta, V., Longo, S., Mistretta, M., and Parisi, M.L. (2020). Review on life cycle assessment of solar photovoltaic panels. *Energies* 13, 252.

Nam, K.T., Kim, D.W., Yoo, P.J., Chiang, C.Y., Meethong, N., Hammond, P.T., Chiang, Y.M., and Belcher, A.M. (2006). Virus-enabled synthesis and assembly of nanowires for lithium ion battery electrodes. *Science* 312, 885–888.

Nam, J.Y., and Logan, B.E. (2011). Enhanced hydrogen generation using a saline catholyte in a two chamber microbial electrolysis cell. *International Journal of Hydrogen Energy* 36, 15105–15110.

Nayeripour, M. and Kheshti, M. (2012). *Sustainable Growth and Applications in Renewable Energy Sources*. InTech, Rijeka.

Nazeeruddin, M.K., Baranoff, E., and Grätzel, M. (2011). Dye-sensitized solar cells: A brief overview. *Solar Energy* 85, 1172–1178.

Ness, D. (2020). Growth in floor area: The blind spot in cutting carbon. *Emerald Open Research* 2, 2.

Pacca, S., Sivaraman, D., and Keoleian, G.A. (2007). Parameters affecting the life cycle performance of PV technologies and systems. *Energy Policy* 35, 3316–3326.

Pappas S. (2016). Facts about silicon. https://www.livescience.com/28893-silicon.html (accessed: Oct 2023).

Parisi, M.L., Maranghi, S., and Basosi, R. (2014). The evolution of the dye sensitized solar cells from Grätzel prototype to up-scaled solar applications: A life cycle assessment approach. *Renewable and Sustainable Energy Reviews* 39, 124–138.

Park, S.K. (2018). Social bonds for sustainable development: A human rights perspective on impact investing. *Business and Human Rights Journal* 3, 233–255.

Parris, T.M., and Kates, R.W. (2003). Characterizing and measuring sustainable development. *Annual Review of Environment and Resources* 28, 559–586.

Peng, J., Lu, L., and Yang, H. (2013). Review on life cycle assessment of energy payback and greenhouse gas emission of solar photovoltaic systems. *Renewable and Sustainable Energy Reviews* 19, 255–274.

Rantik, M. (1999). Life cycle assessment of five batteries for electric vehicles under different charging regimes. *Chalmers University of Technology*, Goteborg, Sweden

Ratner, M. (2004). The physics of solar cells ; third generation photovoltaics: Advanced solar energy conversion. *Physics Today* 57, 71–72.

Reddy, T. A. (2017). *Energy Conversion*, 2nd edn, CRC Press, Boca Raton.

Rozendal, R.A., Hamelers, H.V.M., and Buisman, C.J.N. (2006). Effects of membrane cation transport on pH and microbial fuel cell performance. *Environmental Science and Technology* 40, 5206–5211.

Rydh, C.J., and Sandén, B.A. (2005). Energy analysis of batteries in photovoltaic systems. Part I: Performance and energy requirements. *Energy Conversion and Management* 46, 1957–1979.

Rydh, C.J. (2019). Environmental assessment of NiCd: Battery manufacturing -past and present trends. Linnaeus Eco-Tech 461-470.

Serrano, E., Rus, G., and García-Martínez, J. (2009). Nanotechnology for sustainable energy. *Renewable and Sustainable Energy Reviews* 13, 2373–2384.

Sherif, S.A., Barbir, F., and Veziroglu, T.N. (2005). Wind energy and the hydrogen economy-review of the technology. *Solar Energy* 78, 647–660.

Singh, A., Olsen, S.I., and Pant, D. (2013). Importance of life cycle assessment of renewable energy sources. In: *Green Energy and Technology*, ed. Singh, A., Pant,D., and Olsen, S.I., Springer, New York, pp. 1–11.

Solow, R. (1993). An almost practical step toward sustainability. *Resources Policy* 19, 162–172.

Song, J.H., Murphy, R.J., Narayan, R., and Davies, G.B.H. (2009). Biodegradable and compostable alternatives to conventional plastics. *Philosophical Transactions of the Royal Society B: Biological Sciences* 364, 2127–2139.

Sørensen, B. (2017). *Renewable Energy: Physics, Engineering, Environmental Impacts, Economics and Planning*, 5th edn, Elsevier, Amsterdam.

Speers, A.M., and Reguera, G. (2012). Electron donors supporting growth and electroactivity of geobacter sulfurreducens anode biofilms. *Applied and Environmental Microbiology* 78, 437–444.

Steinbusch, K.J.J., Hamelers, H.V.M., Schaap, J.D., Kampman, C., and Buisman, C.J.N. (2010). Bioelectrochemical ethanol production through mediated acetate reduction by mixed cultures. *Environmental Science and Technology* 44, 513–517.

Stroyuk, O. (2018). Lead-free hybrid perovskites for photovoltaics. *Beilstein Journal of Nanotechnology* 9, 2209–2235.

Stoytcheva, M. (2012). *Pesticides: Formulations, Effects, Fate.* InTech, Rijeka.

Sullivan, J L, Gaines, L, and Energy Systems. (2010). A review of battery life-cycle analysis: State of knowledge and critical needs. US Department of Energy, Oak Ridge.

Sullivan, J.L., and Gaines, L. (2012). Status of life cycle inventories for batteries. *Energy Conversion and Management* 58, 134–148.

Uriarte, F.A. (2010). *Biofuels from Plant Oils: A Book for Practitioners and Professionals Involved in Biofuels, to Promote a Better and More Accurate Understanding of the Nature, Production, and Use of Biofuels from Plant Oils.* ASEAN Foundation, Jakarta.

Wang, C., Waje, M., Wang, X., Tang, J.M., Haddon, R.C., and Yan, Y. (2004). Proton exchange membrane fuel cells with carbon nanotube based electrodes. *Nano Letters* 4, 345–348.

Wee, G., Salim, T., Lam, Y.M., Mhaisalkar, S.G., and Srinivasan, M. (2011). Printable photo-supercapacitor using single-walled carbon nanotubes. *Energy and Environmental Science* 4, 413–416.

Wong, J.H., Royapoor, M., and Chan, C.W. (2016). Review of life cycle analyses and embodied energy requirements of single-crystalline and multi-crystalline silicon photovoltaic systems. *Renewable and Sustainable Energy Reviews* 58, 608–618.

Xu, J., Chen, Y., and Dai, L. (2015). Efficiently photo-charging lithium-ion battery by perovskite solar cell. *Nature Communications* 6, 8103.

Yan, N.F., Li, G.R., and Gao, X.P. (2014). Electroactive organic compounds as anode-active materials for solar rechargeable redox flow battery in dual-phase electrolytes. *Journal of The Electrochemical Society* 161, A736–A741.

Zhang, X., Cheng, S., Huang, X., and Logan, B.E. (2010). The use of nylon and glass fiber filter separators with different pore sizes in air-cathode single-chamber microbial fuel cells. *Energy and Environmental Science* 3, 659–664.

Index

absorbance 104, 107
acidic oxidation 109, 110, 113
activated carbon 100
additives 65, 71, 75, 76, 98, 180
alkaline fuel cell 43
anode 39, 40, 44, 45, 46, 47, 48, 49, 50, 51, 52, 61, 66, 67, 104, 113, 127, 128, 132, 134, 135, 136, 137, 139, 141, 147, 148, 149, 150, 171, 180, 181
arc-discharge 112, 113
automotive 55, 56, 57, 180

barium metastannate 139, 145
battery 40, 45, 52, 54, 57, 58, 65, 68, 72, 82, 106, 121, 164, 167, 170, 171, 180
bioenergy 164
biofuel 52, 165
biogas 165
biomass 114, 115, 118, 119, 120, 159, 164, 165

cadmium metastannate 139, 147
cadmium telluride 12, 23, 168, 172, 173
calcium metastannate 139, 142
capacitance 97, 99, 100, 101, 102, 103, 104, 105, 116, 118, 119, 120, 121
capacitor 53, 98, 99, 100, 101, 103, 104, 105, 106, 145, 164
carbon adatoms 32, 34, 35
carbon black 47, 48, 137
carbon capture 7, 98
carbon dioxide 7, 45, 46, 49, 165
carbon quantum dots 97, 106
carbonization 109, 115, 118, 119, 136
catalyst 39, 40, 45, 47, 48, 52, 53, 55, 131, 180
cathode 39, 40, 44, 45, 46, 47, 48, 49, 50, 51, 52, 61, 67, 104, 112, 113, 127, 139, 141, 171, 181
charge carriers 16, 17, 18, 24, 26, 29, 34, 61, 68, 70, 71, 75, 79, 82, 83, 86, 111
charge storage 98, 104, 105

charging 64, 99, 100, 101, 102, 103, 104, 105, 145, 149, 163, 164, 171
chemical energy 5, 39, 40, 45, 52, 127, 179
cobalt metastannate 139, 145
coffee ground 114, 116, 120
composite polymer electrolyte 64, 84
conducting polymer 39, 67, 104, 105, 106, 134, 141
conductive 63, 70, 97, 98, 99, 178
copolymer 65, 71, 72, 73, 79, 82, 86

direct methanol fuel cell 49, 180
discharging 64, 99, 101, 102, 103, 145, 149, 163
dye-sensitized solar cell 67, 86, 172, 175

electric double-layer capacitor (EDLC) 97, 98, 99, 100, 101, 103, 104, 105, 106, 116, 119, 120
electrical energy 5, 20, 39, 44, 52, 100, 127
electrochemical device 39, 49, 61, 63, 64, 65, 67, 79, 86
electrode 5, 20, 27, 30, 39, 40, 41, 43, 44, 45, 46, 47, 48, 49, 50, 52, 61, 62, 63, 64, 65, 66, 67, 68, 74, 76, 77, 81, 97, 98, 99, 100, 101, 103, 104, 105, 106, 112, 113, 116, 117, 118, 119, 120, 121, 122, 127, 129, 133, 134, 135, 137, 141, 142, 150, 163, 164, 171, 180, 181
electroluminescence 14, 108
electrolyte 39, 41, 43, 44, 45, 46, 47, 48, 49, 51, 52, 61, 62, 63, 64, 65, 66, 67, 68, 69, 70, 71, 72, 73, 74, 75, 76, 77, 78, 79, 80, 81, 82, 83, 84, 85, 86, 87, 88, 97, 98, 99, 100, 101, 103, 104, 105, 116, 117, 118, 120, 121, 127, 132, 135, 138, 141, 163, 169, 171, 175, 180, 181
electrolyzers 46
electron transporting layer 21
energy conversion 4, 5, 6, 7, 8, 10, 13, 14, 24, 40, 42, 43, 55, 63, 161, 163, 164, 166, 179

energy density 7, 50, 57, 79, 82, 97, 101,
    105, 118, 121, 128, 164, 180
energy issue 1, 2, 3
energy policies 2
energy storage 4, 7, 8, 9, 57, 61, 64, 82, 97,
    98, 99, 100, 101, 103, 104, 105,
    116, 118, 119, 121, 122, 161,
    163, 164

filler 65, 66, 67, 71, 75, 80, 81, 82,
    83, 85
fossil fuels 1, 2, 3, 13, 20, 39, 98, 159
fuel cell 4, 20, 39, 40, 41, 42, 43, 45, 47,
    48, 49, 51, 52, 53, 54, 55, 56, 57,
    58, 61, 67, 68, 87, 179, 180

gallium arsenide 22
gallium nitride 22
galvanostatic charge-discharge (GCD) 101,
    102, 116, 117, 118, 119
gamma irradiation 74
gel polymer electrolyte (GPE) 64, 66, 67,
    74, 75, 76, 79, 87
graphene 9, 28, 29, 30, 31, 32, 33, 34, 35,
    36, 37, 120, 131, 138, 139, 141,
    147, 148, 150

hole transporting layer (HTL) 21, 27
hybrid capacitor 97, 98, 104, 105, 106, 120
hydrogen 4, 7, 9, 39, 40, 42, 43, 44, 45, 46,
    47, 49, 51, 52, 53, 54, 55, 56, 57,
    74, 82, 83, 87, 110, 179, 180
hydrothermal 106, 108, 109, 113, 114, 115,
    118, 120, 131, 133, 135, 136, 137,
    140, 142, 143, 144, 145, 146,
    147, 150

illumination condition 16, 17
ionic conduction 64, 65, 68, 70, 71, 81, 85
ionic conductivity 63, 64, 65, 66, 67, 70,
    71, 72, 73, 74, 75, 76, 77, 78, 79,
    80, 82, 83, 84, 85, 86, 87, 117
ionic liquid 65, 71, 75, 77, 78, 79, 80

Jahn–Teller distortion 32

laser ablation 111, 112, 113
lead-acid batteries 170
life cycle analysis 166, 167, 171
line defects 32, 33
liquid crystal polymer electrolytes 85

liquid electrolyte 62, 63, 64, 66, 82,
    87, 100
lithium stannate (lithium tin oxide) 139
lithium-ion batteries 61, 64, 74, 164,
    171, 180

materials science 6, 7, 8, 150
mechanism 4, 14, 17, 27, 39, 40, 47, 48, 54,
    63, 67, 68, 69, 70, 71, 81, 86, 98,
    99, 103, 104, 105, 116, 122, 130
methane 45, 165
microwave pyrolysis 110, 113
mix salt system 75
molten carbonate fuel cell 43, 49
multiple-vacancy defects 32, 34

nanorods 134, 135, 136, 137, 138, 142, 143,
    144, 145
nanosheets 120, 131, 133, 134, 136,
    137, 148
nanotubes 31, 81, 112, 113, 134, 135, 136,
    142, 143, 147, 181
nanowires 134, 135, 142, 144, 149, 181

ocean energy 165
open circuit voltage 82

perovskite solar cell 21, 163, 175
phosphoric acid fuel cell 43, 47
photocatalytic 145, 163, 164
photocurrent 14, 16, 18, 27
photoluminescence 97, 108
photovoltage 14, 15, 16, 18
photovoltaic 5, 6, 12, 13, 18, 21, 22, 23, 53,
    86, 163, 164, 167, 168, 172
plasticizer 65, 66, 71, 75, 76, 77, 85
platinum 39, 40, 43, 45, 46, 47, 48, 52, 53,
    175, 180
P-N junction 15, 16
polarization 45, 46
polymer electrolyte membrane fuel cell
    (PEMFC) 39, 40, 42, 43, 51, 52,
    53, 54, 55, 56, 57, 58, 87, 180
polymer matrix 66, 69, 72, 76, 77, 79, 81,
    82, 86
polymerization 72, 73, 86
porous carbon 40, 118, 133
potassium hydroxide 43, 44, 146
power conversion efficiency 21, 86
power density 46, 49, 52, 55, 57, 82, 87, 97,
    118, 121, 180

power generation 5, 7, 9, 13, 53, 54, 57, 104
power plants 7, 13, 46, 49, 53, 54
pseudocapacitor 97, 98, 104, 105, 120

quantum dots 22, 97, 98, 106
quantum efficiency 14
quasi-neutral region 16

recombination 16, 17, 18, 21, 24, 27, 30, 37, 40, 104
recycling 121, 162, 169, 177, 178, 179
renewable energy 1, 2, 3, 4, 7, 8, 9, 12, 13, 20, 39, 63, 99, 165

Shockley-Queisser limit 22
short-circuit current density 86
silicon solar cell 21, 25, 163, 172
single-vacancy defects 32, 35
SnO$_2$/carbon composites 137
sodium-sulphur batteries 171
solar energy 4, 5, 6, 9, 18, 20, 24, 163, 164
solid electrolyte 62, 104
solid oxide fuel cell 43, 45
solid polymer electrolytes 63, 64, 180
Stone–Wales defects 32, 33
strontium metastannate 139, 142

supercapacitors 28, 61, 68, 87, 97, 98, 99, 104, 117, 164
surface area 29, 48, 49, 67, 81, 97, 98, 99, 100, 104, 114, 116, 118, 120, 131, 134, 136, 138, 139
sustainable development 159, 160, 161, 162

technology 1, 2, 12, 13, 18, 20, 23, 24, 31, 39, 46, 52, 55, 57, 61, 105, 112, 118, 159, 161, 165, 168, 178, 180, 181
ternary tin oxides 138
thermodynamics 4, 5, 40, 42, 43
thin-film solar cells 168, 172, 173, 178
tidal power 166
tin (II) oxide 130
tin (IV) oxide 132
tin-based alloys 129
tin-based oxides 128, 130, 139

vacancy healing 35, 36
vacancy mechanism 68

wave energy 166
work 4, 5, 40, 41, 42, 43

zinc metastannate 139, 148
zinc stannate 148

For Product Safety Concerns and Information please contact our EU
representative GPSR@taylorandfrancis.com
Taylor & Francis Verlag GmbH, Kaufingerstraße 24, 80331 München, Germany

www.ingramcontent.com/pod-product-compliance
Lightning Source LLC
Chambersburg PA
CBHW070717220326
41598CB00024BA/3198